AF655736

MEDICINAL PLANTS
OF MADHYA PRADESH

MEDICINAL PLANTS
OF MADHYA PRADESH

Dinesh Jadhav
Assistant Professor
Department of Botany
Govt Arts and Science College, Ratlam (M.P.)

AUTHORSPRESS

Worldwide Circulation through Authorspress Global Network
First Published in 2022
by
Authorspress
Q-2A Hauz Khas Enclave, New Delhi-110 016 (India)
Phone: (0) 9818049852
E-mail: authorspressgroup@gmail.com
Website: www.authorspressbooks.com

Medicinal Plants of Madhya Pradesh
ISBN 978-93-5529-112-7

Copyright © 2022 Dinesh Jadhav

All rights reserved. No part of this book may be reproduced or transmitted in any form or by any means, electronic, mechanical, photocopying, or otherwise, without the express written consent of the author.

Printed in India at Thomson Press (India) Limited

Dedicated to
The Memory of My Grandmother
Late Mrs. Kanchan Devi Jadhav

Acknowledgements

First and foremost I express my deep sense of gratitude and indebtedness to Respected Dr R. L. Mehra, Chairman, Madhya Pradesh Public Service Commission, Indore (M.P.), Respected Dr. Anil K. Goel, Deputy Director, Botanic Garden, National Botanical Research institute (NBRI), Lucknow (Uttar Pradesh), Dr. S. K. Agarwal (Retd Professor, Department of Botany, DBS College, Dehra Dun, Uttarakhand) and Dr R.L.S. Sikarwar (Deendayal Research Institute, Chitrakoot, Satna) for encouragement, keen interest, support and co-operation.

I express my deep sense of gratitude to Respected Shri. J. C. Chauhan, Retd. Prof. and Head, Department of Botany, Govt. Arts and Science College, Ratlam (M. P.), Respected Dr. Sanjay Wate, Principal, Govt. Arts and Science College, Ratlam (M. P.), Respected Dr. Rajkumar Katare, Principal, Govt. Girls Degree College, Ratlam (M. P.), Respected Mrs. Sushma Raj Katare, Professor and Head, Department of Botany, Govt. Girls Degree College, Ratlam (M. P.), Respected Dr. Abhay Pathak, Professor, Govt. Vivekananda Commerce College, Ratlam (M. P.), Respected Dr Bhavana Deshpande, Professor, Department of Mathematics, Govt. Arts and Science College, Ratlam (M. P.) and Respected Shri. Manoj Verma, Professor, Department of English, Govt. College, Satna (M. P.) for encouragement, keen interest, support and valuable co operation.

Dinesh Jadhav

Preface

Medicinal plants belong to the earliest known health care products that have been used by the mankind. In India, the earliest reference to the medicinal value of plants appears in *Rigveda*, in which a brief reference to the healing property of plants has been made. However, it is in the *Ayurveda* that definite properties and uses of drugs and drug yielding plants have been discussed. In fact, Ayurvedic medicine had its origin from the works of *Ayurveda*. The later works of *Charak* and *Susruta*, namely *Charak-Samhita* and *Susruta Samhita* added invaluable knowledge to the science of medicinal plants.

Most of the drug plants are found in the tropics growing in wild condition and are mainly used by herb doctors and Ayurvedic vaidyas, who refer to them as *jari-butis*. In the modern age with the fast emergence of western modern medicine and synthetic drugs, the importance of plants and herbs started declining as more and more attention was paid to the development of synthetic drugs. However, with these synthetic drug came the problems of side effects and other clinical complications; the reason in the increasing awareness about the limitations of the synthetic drugs. Because of natural herbal drugs are safe and effective; now herbal medicine and natural products are in big demand all over the World.

This book is an effort to put together the vast variety of medicinal plants found in Madhya Pradesh. This book well helps in generating an interest in *Ayurveda* and medicinal plants of Madhya Pradesh.

I hope that this compilation of most commonly used medicinal plants will be of use to Ayurvedic practitioners, students and researchers as well as to medical and scientific communities worked on the medicinal plants of Madhya Pradesh.

Dinesh Jadhav

Contents

Introduction

Man has depended on plants from time immemorial; plants have been a rich source for therapeutics. Medicinal plants belong to the earliest known health care products that have been used by the mankind. Over three-quarters of the World population rely on the use of traditional medicines for their primary health care needs. The medicinal plants are not only major components of the many formulations used in indigenous systems of medicine but also of a large number of drugs in Allopathy. In India, with one of the oldest ethnobotanical tradition in the World, medicinal plants are the backbone of all major systems of medicine. One of the earliest references to the Medicinal properties of plants is in the *Rigveda* (3500-1800 BC). *Ayurveda*, the ancient Indian system of medicine based on that medicinal property of plants. There are many more useful medicinal plants has been described under Ayurvedic medicinal system by several authors at that time i.e. '*Charakh*' and '*Susruta*' etc. Their Ayurvedic text manifests that an incredible 1600 of the 2000 old Ayurvedic formulations are plant based. After that there are many more books have been written under heading medicinal plants of India, but they could not fulfil all the requirements of research scholar and readers who worked/interested in the field of Medicinal Plants of Madhya Pradesh. In India, Madhya Pradesh is known to harbour a rich wealth of Medicinal Plants. The State lies in the central region of the Country between latitudes 21^0 6' – 26^0 30' N and longitudes 74^0 9' – 81^0 48' E covers an area of 3,08,245 sq. km. The present book is focused on Medicinal Plants of Madhya Pradesh. In this book the description on each plant species includes Distribution in the State, Botanical description, Chemical constituents, Ayurvedic Medicinal properties, Ayurvedic Medicinal uses and Ayurvedic preperations. This book is helpful for Botanists, NGO's and research workers interested in carrying out the researchers in the field of Ayurveda and Medicinal uses of the plants, BAMS students, the student of Botany and the persons engaged in Pharmaceutical concerns as well as other reader's interested in the field of Herbal-medicine.

1. *Abelmoschus esculentus* (L.) Moench. (Malvaceae)

Syn.: *Hibiscus esculentus L.*

Vern.: Bhindi.

Distribution: Commonly cultivated throughout the State.

Botany: Hispidly hairy herb, with 3 to 7-lobed leaves and yellow flowers with purple centre. Fruits cylendric, 5-angled, beaked. Flowering & Fruiting: April-September.

Medicinal Parts: Root, Fruits, Seeds.

Chemical Constituents: Fruits contain 5-hydroxytryptamin, oxalates, iodine, flourine, Calcium, phosphorus, Magnesium, vitamin-C, fats, carbohydrates, fibres, Amino-acids, Polyphenols.

Medicinal Properties: Fruits – Aphrodisiac, constipating, cooling, demulcent, diuretic, emollient, stomachic.

Seeds – Antispasmodic, aphrodisiac, demulcent, emollient, stimulant.

Medicinal Uses: Fruits – Abdominal disorders, anaemia, diarrhoea, dysuria, gonorrhoea, Pruritus, spermatorrhoea.

Seeds – Gonorrhoea.

2. *Abelmoschus moschatus* (L.) Medic. (Malvaceae)

Syn.: *Hibiscus abelmoschus L.*

Vern.: Kasturi Bhindi, Mushk dana.

Distribution: In mixed forests and along cultivated fields. Mandla.

Botany: Strigose undershrubs, with ovate-cordate leaves. Flowers bright yellow with dark purple centre. Capsules acuminate, villous. Seeds reniform.

Flowering & Fruiting: September – October.

Medicinal Parts: Seeds.

Chemical constituents: Seeds contains Farnesol, ambrettolide, sterols, terpenes and aliphatic compounds.

Seed oil contains 12, 13 – epoxyoleic acid, malvalic acid, sterculic acid & phospholipids.

Medicinal Properties: Seeds – Acrid, antispasmodic, aphrodisiac, aromatic, bitter, carminative, cooling, demulcent, digestive, diuretic, pectoral, stimulant, stomachic, tonic.

Medicinal Uses: Seeds – Dyspepsia, flatulence, gonorrhoea, hysteria, intestinal complaints, itching, leucoderma, nervous debility, snakebite, stomatitis, urinary discharges, vomiting.

3. *Abrus precatorius* L. (Fabaceae)

Syn.: *Glycine abrus L.*

Vern.: Ratti.

Distribution: On roadside trees and shrubs. Balaghat, Bhopal, Damoh, Hoshangabad, Indore, Jabalpur, Mandla, Panna, Raisen, Ratlam, Rewa, Sagar, Satna, Seoni, Shahdol, Shivpuri, Sidhi, W. Nimar.

Botany: Twining shrubs, with 10-20 pairs of leaflets and pink or white flowers having 9 stamens. Pods 3-5 seeded.

Flowering & Fruiting: August-February.

Medicinal Parts: Leaves, Roots, Seeds.

Chemical constituents: Root contains Stigmasterol, Glycine, Glutamic acid, Aspartic acid, Tyrosin, Quercetin, Luteplin, Kaempferol, Taxalbumin; Alkaloid-abrin; Glucoside-abralin; Anthocyanin-abranin.

Seeds contain Lupeol acetate, Sitosterol, Sucrose, Gallic acid, Orietin, Isoorientin, Abrussic acid, Haemagglutinin.

Leaves contain Precol, Abrol, Abrasine, Precasine.

Medicinal Properties: Root – Acrid, alexiteric, aphrodisiac, Bitter, diuretic, emetic.

Leaves – Acrid, alexiteric, aphrodisiac, Bitter, diuretic, emetic.

Seeds – Abortifacient, aphrodisiac, emetic, Purgative.

Fruits – Acrid, aphrodisiac, Bitter.

Medicinal Uses: Root – Asthma, biliousness, colic, fever, gonorrhoea, Jaundice, stomatitis, tuberculous glands.

Leaves – Asthma, cold, colic, cough, fever, gonorrhoea, headache, hoarseness, itching, jaundice, leucoderma, skin diseases, stomatitis, swelling, tuberculous glands.

Seeds – Diarrhoea, dysentery, leprosy, nervous disorders, paralysis, sciatica, ulcer.

4. *Abutilon indicum* (L.) Sweet (Malvaceae)

Syn.: *Abutilon asiaticum* (L.) Sweet.; *Sida guineensis* Schum.

Vern.: Kanghi, Mudra.

Distribution: In Wastelands, Stream sides and Hill slopes. Balaghat, Bhopal, Damoh, Indore, Panna, Satna, Shahdol, Sidhi.

Botany: Undershrubs, with ovate – suborbicular leaves and axillary, solitary flowers. Fruits with 15-25 mericarps.

Flowering & Fruiting: September – January.

Medicinal Parts: Root bark, Leaves, Seeds.

Chemical constituents: Plant contains Gossypetin, Glucosides, Cyanidin, Rutinosides, essential oil.

Root contains Fatty acids, Aspargin, Rafinose.

Stem & Leaves contain Alkanol, Beta–Sitosterol, Vanillic acid, Caffeic acid, Fumaric acid, Fructose, Glucose, Galactose, Leucin, Histidine, Threonine, Serine, Glutamic acid and Aspartic acid.

Medicinal Properties: Root – Antipyretic.

Root bark – Astringent, diuretic.

Leaves – Demulcent, diuretic, emollient.

Seed – Aphrodisiac, demulcent, laxative.

Medicinal Uses: Root – Eye diseases, fever, leprosy, leucoderma, Piles, stone in bladder, strangury, uterine hemorrhagic discharges.

Root bark – Strangury, urinary complaints.

Leaves – Boils, bronchitis, diarrhoea, fever, gonorrhoea, gums, inflammation, lumbago, piles, toothache, ulcers, vaginal infections, wounds.

Seeds – Bronchitis, chronic chest trouble, cough, cystitis, gleet, gonorrhoea, piles.

Ayurvedic preperations: *Atibala taila, Chavyanprash, Atibala ghrit, Mahanarayan taila, Mahavishgarbh taila, Atibala svarasa.*

5. *Acacia arabica auct.* (Mimosaceae)

Syn.: *A. nilotica* (L.) Willd. ex Delile ssp. *indica* (Benth.) Brenan.

Vern.: Babool.

Distribution: In dry deciduous forests and roadsides. Balaghat, Bhind, Bhopal, Chhatarpur, Damoh, Gwalior, Hosahangabad, Indore, Mandla, Panna, Raisen, Sagar, Satna, Seoni, Shivpuri, Sidhi.

Botany: Trees, with long, white stipular spines. Pinnae 2-11 pairs and leaflets 7-25 pairs. Heads bright yellow, axillary, pedunculate, pods moniliform.

Flowering & Fruiting: July – January.

Medicinal Parts: Stem, leaves, fruits, stem bark, gum, seeds.

Chemical constituents: Bark contains Beta-amyrin, Gallic acid, tannins, catechin, quercetin, leucocyanidin.

Fruit contains Gallic Acid, Catechin, Chlorogenic acid.

Gum contains Galactose, Arabic Acid, Calcium, Magnesium.

Seeds contains Ascorbic acid, Niacin, Thiamine, Amino acids.

Medicinal Properties: Gum – Antidysenteric, astringent, styptic, tonic.

Stem bark – Acrid, alexipharmic, anthelmintic, aphrodisiac, astringent, demulcent, emetic.

Fruit – Acrid, astringent, cooling, expectorant, sweet.

Leaves – Antipyretic, astringent.

Medicinal Uses: Gum – Biliousness, bone fractures, diabetes, diarrhoea, dysentery, leprosy, sore throat, urinary discharges.

Stem bark – Biliousness, bronchitis, burning sensation, cough, diarrhoea, dysentery, leucoderma, piles, urinary discharges.

Fruit – Biliousness.

Leaves – bone fractures, bronchitis, eye diseases, gonorrhoea, piles.

Ayurvedic preperations: *Babularishta, Lavangadi vati, Babbula churna, Lavangarista.*

6. *Acacia catechu* Willd. (Mimosaceae)

Syn.: *Mimosa catechu* L. f.; *M. catechuoides* Roxb.

Vern.: Khair.

Distribution: In dry deciduous forests. Balaghat, Bhopal, Chhindwada, E. Nimarh, Gwalior, Hosahangabad, Indore, Jabalpur, Mandla, Panna, Raisen, Sagar, Satna, Shivpuri, Sidhi.

Botany: Trees; all part pubescent; prickles hooked. Leaves 2-pinnate, with glands on the petiole and between the pinnae; leaflets 10-30 pairs. Spikes pale yellow. Pods flat.

Flowering & Fruiting: May – September.

Medicinal Parts: Stem bark, resinous extract of heart wood.

Chemical constituents: Inner wood contains Catechin & Catechutannic acid (Catechutannic acid contain about 50% Tannin).

Medicinal Properties: Stem bark – Acrid, anthelmintic, antidysenteric, antipyretic, astringent, bitter, cooling.

Resinous extract – Acrid, anthelmintic, aphrodisiac, astringent, bitter, hot, tonic.

Medicinal Uses: Stem bark – Anaemia, boils, bronchitis, cough, diarrhoea, diseases of the tooth, elephantiasis, erysipelas, heaviness, indigestion, inflammation, itching, leprosy, leucoderma, psoriasis, sore throat, skin diseases, ulcers, urinary discharges, worms.

Resinous extract – Cough, diarrhoea, leprosy, leucoderma, leucorrhoea, oedema, piles, throat diseases, toothache, ulcer, urinary diseases, vaginal discharges.

Ayurvedic preperations: *Khadirarista, Khadirsar, Khadiradi vati, Khadiradi kwath, Khadirastak, Kutajarista, Lavangadi vati, Chandanadi vati, Vrihtkhadiradi vati.*

7. *Acacia farnesiana Willd.* (Mimosaceae)

Vern.: Bilayati Babool.

Distribution: Balaghat, Damoh, Indore, Jabalpur, Rewa, Raisen, Shivpuri.

Botany: Tree, grows upto 7-8 m. height covered with small thorns; Paripinnate compound leaf; Yellow coloured small flowers; Head inflorescence; fruit legume.

Flowering & Fruiting: September-March.

Medicinal Parts: Bark.

Chemical Constituents: Flowers are the source of Cassie Perfume. Pods contain tannin. Seeds contain a Fatty oil.

Medicinal Properties: Bark – Acrid, Alexetric, Anthelmintic, Antidysenteric, Hot.

Medicinal Uses: Gum – Aphrodisiac, Sweet, Tonic.

Bark – Bronchitis, Diseases of Mouth & Blood, Erysipelas, Excessive haemorrhage, Inflammation, Itching, Leucoderma, Stomatitis, Ulcers.

8. *Acacia leucophloea* (Roxb.) Willd. (Mimosaceae)

Syn.: *A. alba* Willd; *Mimosa leucophloea* Roxb.

Vern.: Safed kikar.

Distribution: In wastelands and roadsides. Balaghat, Bhopal, Bhind, Dhar, Gwalior, Hosahangabad, Indore, Mandla, Morena, Panna, Raisen, Ratlam, Sagar, Satna, Seoni, Shivpuri, Sidhi, Tikamgarh.

Botany: Trees, with stipular-spines in pairs. Pinnae 3-15 pairs with concave glands in between and 5-30 pairs of leaflets. Heads pale-yellow, in terminal panicles. Pods flat, linear.

Flowering & Fruiting: May-March.

Medicinal Parts: Bark.

Chemical constituents: Stem Bark contain n-hexacosanol, β-amyrin, β-sitosterol, tannin, isookannin, cyanin, leuoxol.

Heartwood contain n-hexacosanol, β-amyrin.

Medicinal Properties: Bark – alexeteric, anthelmintic, antipyretic, astringent, bitter, constipating, demulcent, expectorant, styptic, thermogenic, vulnerary.

Medicinal Uses: Bark – Bronchitis, cough, diarrhoea, dysentery, erysipelas, fever, inflammation, leprosy, leucoderma, pruritus, skin diseases, ulcers, vomiting, wounds.

9. *Acacia pennata* (L.) Willd. (Mimosaceae)

Syn.: *Mimosa pennata* L.

Vern.: Chundria.

Distribution: In mixed dry deciduous forests. Balaghat, Damoh, Hosahangabad, Indore, Mandla, Panna, Raisen, Rewa, Sagar, Seoni, Sidhi.

Botany: Large, scandent bush, with polished bark. Prickles slightly recurved. Pinnae 8-15 pairs, with glands between the upper ones; leaflets 40-50 pairs. Flowers creamy-white or pale-yellow. Pods stalked, linear-oblong, glabrous.

Flowering & Fruiting: April-October.

Medicinal Parts: Roots.

Chemical Constituents: Bark contains lupeol and α-Spinasterol and Stem contain β-Sitosterol.

Medicinal Properties: Bark-Acrid, Bitter, Sour.

Medicinal Uses: Bark-Asthma, Biliousness, Bronchitis, Diseases of Blood.

10. *Acacia sinuata* (Lour.) Merr. (Mimosaceae)

Syn.: *Mimosa sinuata* Lour.; *M. concinna* Willd.; *Acacia rugata* (Lamk.) Merr.; *A. Concinna* (Willd.) DC.

Vern.: Shikakai.

Distribution: Jabalpur, Sagar.

Botany: A scandent spreading spiny shrub; leaves pinnately compound; flowers in globose head, yellow-whitish or pink; fruits fleshy, beaked & constricted.

Flowering & Fruiting: March January.

Medicinal Parts: Pods.

Chemical constituents: Leaves contain Tartaric, citric, oxalic, succinic & ascorbic acids, tartaric racimase, rutin, tannin along with alkaloids-colycotomine & nicotine.

Pods contain Saponins.

Seeds contain Saponins which possesses β-amrine, acacinin A & B, Carbohydrate glucose, arabinose, xylose, fucose & rhamnose.

Medicinal Properties: Pods – Anthelmintic, astringent, bitter, cooling, depurative, detergent, diuretic, emetic.

Medicinal Uses: Pods – Burning sensation, constipation, eczema, erysipelas, leprosy, leucoderma, strangury.

11. *Acalypha indica L.* (Euphorbiaceae)

Syn.: *Acalypha spicata; A. ciliata.*

Vern.: Kuppi.

Distribution: Common as weed in gardens, fields, old buildings and vacant plots near villages. Chhatarpur, Dhar, Indore.

Botany: Erect herbs, with ovate or ovate – rhomboid leaves. Flowers greenish, in androgynous spikes; bracts concealing the capsules, Shortly dentate, capsules hispid.

Flowering & Fruiting: July-November.

Medicinal Parts: Whole plant.

Chemical constituents: Plant contains Kaempferol, Acalyphamide, sterols, cyanogenic glyceride.

Medicinal Properties: Whole plant – Anthelmintic, carminative, diuretic, emetic, expectorant, laxative.

Medicinal Uses: Whole Plant – Asthma, arthritis, bronchitis, burns, earache, pneumonia, ringworm, scabies, skin diseases, snake-bite, toothache.

12. *Achyranthes aspera L.* (Amaranthaceae)

Vern.: Apamarg.

Distribution: Common in wastelands. Balaghat, Bhopal, Chhatarpur, Damoh, Jabalpur, Narsimhapur, Panna, Rewa, Satna, Seoni, Sidhi, Tikamgarh.

Botany: Erect hairy herbs, with oval–round pubescent – villous leaves. Flowers rather plump in appearance, deflexed, spinescent, in lax spikes.

Flowering & Fruiting: August – February.

Medicinal Parts: Whole plant, Leaves, Roots, Seeds.

Chemical constituents: Seed contains Saponin A & B, Ecdy steron, oligosaccharide.

Root contains Oleonilic acid, maslinic acid.

Unripe fruit contains Saponin C & D.

Plant contains Nitrogenous compound – Achranthine & alkaloids.

Medicinal Properties: The plant is acrid, anodyne, anthelmintic, anti-inflammatory, bitter, carminative, demulcent, depurative, digestive, diuretic, stomachic, sudorific.

Medicinal Uses: Root – Abdominal pain, blood diseases, cancer, colic, deafness, disease of the tooth, heart diseases, itching, leucoderma, liver disorders, stone in bladder, tumor, vomiting, wound.

Leaves – Gonorrhoea, leprosy.

Seeds – Cold, colic, cough, hydrophobia, leprosy, scabies, toothache.

Ayurvedic preperations: *Apamargkshar tila, Apamarg asav, Apamarg avaleh, Cystone, Kushta sam-ul-far, kushta shangraf, Kushta hartal warqi.*

13. *Adansonia digitata* L. (Bombacaceae)

Vern.: Gorakh-imli.

Distribution: *A native of tropical Africa, planted in various parts of M. P. Bhopal, Dhar, Indore.*

Botany: Deciduous trees, With digitately 3 to 9-foliolate leaves and white, pendulous flowers. Fruits ellipsoid, woody.

Flowering & Fruiting: March – May.

Medicinal Parts: Stem bark, fruits & leaves.

Chemical constituents: Pulp & fruit contains Free tarteric acid, potassium bitartrate.

Leaves contain Mucilage.

Stem bark contains Bitter crystalloid – Adansenin.

Medicinal Properties: Leaves – Astringent, febrifuge, sudorific, tonic.

Stem bark – Astringent, cooling, stomachic.

Fruit pulp – Aperient, appetiser, astringent, cooling, demulcent.

Medicinal Uses: Leaves – Burning sensation, earache, inflammation, ophthalmia, swelling.

Stem bark – Intermittent fever.

Fruit pulp – Bronchial asthma, diarrhoea, dysentery, haemoptysis, itching, urticaria.

Ayurvedic preperations: *Goraksacincavaleha.*

14. *Adenanthera pavonia L.* (Mimosaceae)

Vern.: Rakta-kanchan.

Distribution: Planted in gardens. Bhopal.

Botany: A medium sized, handsome deciduous tree. Flowers in short-peduncled spike-like racemes, axillary or panicled at the ends of the branches. Corolla pale-yellow. Pods flat, falcate, 8-15 seeded.

Flowering & Fruiting: March-August.

Medicinal Parts: Bark, Leaves, Seeds, heartwood.

Chemical constituents: Bark conatains Octacosanol, dulcitol, glucosides of β-sitosterol and stigmasterol.

Seed contains Stigmasterol and its glucoside, dulcitol, polysaccharides; non-protein amino acids as γ-methylene glutamic acid, γ-methlene fatty acids as palmitic, stearic, arachidic, lingoceric, oleic, linoleic acid.

Heartwood contains Srobinetin, chalcone, butein, ampelopsin and dihydrorobinetin.

Medicinal Properties: Bark & Leaves – Aphrodisiac, astringent, vulnerary.

Seeds – Antiemetic, aphrodisiac, astringent, bitter, cooling, febrifuge, sweet.

Heart wood – Aphrodisiac, astringent, haemostatic.

Medicinal Uses: Bark & Leaves – Gout, ulcers.

Seeds – Burning sensation, fever, gout, vomiting.

Heart wood – Dysentery, Haemorrhages.

15 *Adhatoda zeylanica* Medic. (Acanthaceae)

Syn.: *A. vasica* Nees.; *Adennanthera vasica*; *Justicia adhatoda* L.

Vern.: Adusa.

Distribution: Common in fields and wastelands. Chhatarpur, Damoh, Hosahangabad, Indore, Rewa, Satna, Seoni.

Botany: Shrubs, with elliptic leaves. Flowers white or creamish – white in elongate spikes; corolla – tube mottled. Capsules 4 – seeded. *Flowering & Fruiting:* December-March.

Medicinal Parts: Whole plant – Leaves, flowers, root bark.

Chemical constituents: Leaves contains A bitter vasicine, Adhatodic acid, volatile oil, vasa, Resin, gum, sugar & yellow pigment.

Plant contains Alkaloids (Vasicoline, Adhatodine, Vasicolinine, Anisolin), Vasicine & Vasicinone.

Medicinal Properties: The plant is acrid, anti-inflammatory, anthelmintic, antileprotic, antiperiodic, antiseptic, antispasmodic bitter, bronchodilator, cooling, diuretic, expectorant, pungent, stimulant.

Medicinal Uses: Asthma, blood impurities, chronic bronchitis, cold, cough, diseases of the mouth, fever, gonorrhoea, haemorrhagic diseases, heart troubles, jaundice, leprosy, ophthalmia, respiratory disorders, rheumatism, skin diseases, tumors, urinary diseases, whooping cough.

Ayurvedic preperations: *Vasaka churna, Vasaka kwath, Vasakarista, Vasaka ghrita, Vasa avaleha, Vasa kshar, Vasa chandanadi ghrit, Vasadi kwath, Phalatrikadi kwath, Kanakasav, Bhunimbadi kwath, Chavyanprash, Geriforte, Shagat aijaz, Bakhor dama, Vasapanaka, Vasaputapaka.*

16. *Aegle marmelos* (L.) Corr. ex Roxb. (Rutaceae)

Vern.: Bilvapatra.

Distribution: Balaghat, Damoh, Dhar, Hosahangabad, Panna, Seoni, Tikamgarh.

Botany: Deciduous trees, armed with spines. Leaves 3-foliolate and flowers creamy – white in axillary panicles. Fruits globose with stony rind and mucilaginous pulp. *Flowering & Fruiting:* March – April.

Medicinal Parts: Whole plant, Roots, bark, leaves, fruits, flowers.

Chemical constituents: Fruits contains Gum, pectins, reducing sugars, sugars, oil, and tannin.

Leaves & Root contains Reducing sugars, Tannins, Iron, Calcium, Magnesium, and Potassium.

Medicinal Properties: Ripe fruit – Acrid, antipyretic, antiscorbutic, appetiser, aromatic, bitter, cooling, febrifuge, nutritive, stimulant, sweet, tonic.

Root – Sweet.

Leaves – Astringent, bitter, digestive, febrifuge, laxative.

Medicinal Uses: Ripe fruit – Diarrhoea, biliousness, dysentery, dyspepsia.

Root – Abdominal pain, colic, fever, dyspepsia, dysuria, nervous disorders, oedema, rheumatism, urinary trouble, vomiting.

Leaves – Deafness, inflammation, ophthalmia.

Ayurvedic preperations: *Bilvapanchak kwath, Bilvadichurna, Bilvadighrit, Bilva taila, Bilvadileh, Brhidgangadhar churna, Bilvamuladi gutika, Anu taila.*

17. *Aerva lanata* (L.) Juss. ex Schultes (Amaranthaceae)

Syn.: *Achyrantuhus lantana* L.; *Celosia lanata* L.

Vern.: Gorakhganja.

Distribution: Common as weed in wastelands, along river banks and roadsides. Balaghat, Betul, Chhatarpur, Damoh, Dhar, E. Nimar, Hosahangabad, Indore, Mandla, Panna, Ratlam, Rewa, Satna, Seoni, Sidhi, Tikamgarh, W. Nimar..

Botany: Perennial herbs, with suborbicular to lanceolate leaves. Flowers bisexual, in axillary spikes passing into long leafy inflorescence. Fruits rotund, compressed.

Flowering & Fruiting: July-April.

Medicinal Parts: Whole plant.

Chemical constituents: α-Amyrin, campesterol, β-sitosterol, palmitic acid, betuline, syringic acid, vanllic acid, feruloyltyramine, aervine, methylervine, aervoside, aervolanine.

Medicinal Properties: Plant – Astringent, bitter, cooling, diuretic, emollient, suppurative, vermifuge.

Medicinal Uses: Plant – Boils, Cough, Strangury.

18. *Agave americana* L. (Agavaceae)

Syn.: *A. cantala* Roxb.

Vern.: **Ketaki.**

Distribution: Cultivated as ornamental in Parks & Gardens. Bhopal.

Botany: Short-Stemmed xerophytic perennial with long narrow leaves. *Flowering & Fruiting:* January-June.

Medicinal parts: Leaf.

Chemical constituents: Saponins.

Medicinal properties: *Root* – Diuretic, Diaphoretic, Anti-syphillitic.

Juice of Leaves – Laxative, Diuretic, Emmenagogue.

Medicinal uses: *Juice of Leaves* – Scurvy, syphilis, Scrofula, Cancer.

19. *Ageratum conyzoides* L. (Asteraceae)

Vern.: Osdi.

Distribution: Common weed of moist places along the irrigation channels and strems. Balaghat, Hoshangabad, Sidhi.

Botany: An annual hairy herb; leaves ovate – cordate, serrate; heads white in terminal corymbs.

Flowering & Fruiting: Round the year.

Medicinal Parts: Leaves.

Chemical constituents: Essential oil, Saponins, Flavonoids, alkaloids, amino acids, sugars, Vitamin – A & B.

Medicinal Properties: Leaf – Styptic.

Root – Antilithic.

Medicinal Uses: Leaf – Eye diseases, Leprosy, tetanus, wound.

Root – Eye diseases.

20. *Ailanthus excelsa* Roxb. (Simaroubaceae)

Vern.: Maharukh.

Distribution: Chhatarpur, Hosahangabad, Indore, Jabalpur, Seoni, Shivpuri, Sidhi.

Botany: Large trees, with pinnate leaves crowned at the end of branches. Flowers yellow, in terminal and axillary panicles. Fruit a samara, spindle shaped, 1-seeded.

Flowering & Fruiting: December – April.

Medicinal Parts: Leaves, bark.

Chemical constituents: Bark contains Glaucarubin, Glauca rubinone, Excelsin, Glaucarubol, Ailanthion, Ailantic acid, Malantahin, Tricontane, Hexatricontane.

Leaves contain Ailantic acid, Beta – sitosterol (vitexin).

Medicinal Properties: Stem bark – Anthelmintic, antidysenteric, aromatic, astringent, bitter, diarrhoea, febrifuge.

Leaves – Bitter, carminative, febrifuge, tonic.

Medicinal Uses: Stem bark – Asthma, bronchitis, chronic fever, diarrhoea, dysentery, dyspepsia, earache, leprosy, piles, skin diseases, vomiting, worm infestation.

21. *Alangium salvifolium* (L.f.) Wang. (Alangiaceae)

Syn.: *A. lamarckii* Thw.; *Grewia salvifolium* (L. f.) Suppl.

Vern.: Ankol.

Distribution: Damoh, Mandla.

Botany: A small tree with spinescent branches and grey bark. Leaves simple, alternate, membranous, varying in size and shape. Flowers in axillary clusters, small, white, and fragrant. Fruits berry, globular, crowned with persistent calyx-limb, purplish-red to black.

Flowering & Fruiting: February– May.

Medicinal Parts: Root, Root bark, Leaves, Fruits, seeds.

Chemical constituents: Flower contain Deoxytubulosine alkaloid.

Fruits contain Cephaeline, deoxytubulosine, alangside, sterol, N-methylcephaeline.

Seed contain Alamarckine, emetin, cephaeline, isocephaeline & other alkaloids.

Leaves contain Alkaloids, sterols & terpenoids.

Stem bark contain Tubulosine, cephaeline, psychotrine alkaloids.

Medicinal Properties: Root – Anthelmintic, laxative.

Root bark – Anthelmintic, astringent, bitter, emetic, pungent, purgative.

Fruit – Astringent, laxative, refrigerant.

Seed – Cooling, tonic.

Medicinal Uses: Root bark – Colic, biliousness, fever, skin diseases.

Leaves – Rheumatic pain.

Fruits – Haemorrhages.

Seed – Boils, Haemorrhages.

22. *Albizia lebbeck* (L.) Willd. (Mimosaceae)

Syn.: *Mimosa lebbeck* L.; *Acacia speciosa* Willd.

Vern.: Sirus.

Distribution: In mixed forests and roadsides. Bhopal, Damoh, Gwalior, E. Nimar, Hosahangabad, Indore, Jabalpur, Panna, Raisen, Sagar, Shivpuri.

Botany: Deciduous trees, with 1-4 pairs of pinnae and 4-10 pairs of leaflets. Flowers pale-green, in umbellate heads. Pods flat, straw-coloured.

Flowering & Fruiting: April – September.

Medicinal Parts: Stem bark, leaves, flowers, and seed.

Chemical constituents: Bark contains Tannins, saponin, Echinocytic acid, and β – sitosterol.

Pods contain Saponins, Echinocytic acid, Olionilic acid, Albigenine & Albigenic acid, lebbekanin – C, Glucose & Rhamnose.

Seed contains Triterpenic saponin.

Medicinal Properties: Stem bark – Acrid, alexiteric, anthelmintic, astringent, bitter, cooling.

Root – Astringent.

Flower – Aphrodisiac, emollient.

Seed – Aphrodisiac, astringent.

Medicinal Uses: Stem bark – Asthma, blood disorders, bronchitis, conjunctivitis, cough, diarrhoea, eczema, erysipelas, helmintic inflammation, itching, leprosy, leucoderma, malaria, migraine, Oedema, paralysis, piles, Rhinitis, sinusitis, skin diseases, snake bite, ulcer, urticaria, worm Infestation.

Root bark – Good for gums.

Leaves – Nightblindness, Ophthalmia.

Seed – Gonorrhoea.

Ayurvedic preperations: *Maha-sirisagad, Sirisarishta, Dasnaga lep.*

23. *Albizia odoratissima* (L. f.) Benth. (Mimosaceae)

Vern.: Chichwa.

Distribution: In mixed deciduous forests. Balaghat, Bhopal, Chhatarpur, Damoh, Gwalior, Hosahangabad, Jabalpur, Mandla, Panna, Satna, Seoni, Shivpuri, Sidhi.

Botany: A medium sized unarmed tree; leaves pinnately compound; flowers white, fragrant, sessile, numerous, in small globose heads; fruits shortly stacked pods.

Flowering & Fruiting: April-June.

Medicinal Parts: Bark.

Chemical constituents: The bark contains tannis composed of d-catechin, isomers of leucocyanidin and melacidin, a leucoanthocyanidin as lebbecacidin, friedelin, β-sitosterol & machaerinic acid.

Medicinal Properties: Bark – Acrid, astringent, cooling, depurative, expectorant.

Medicinal Uses: Bark – Bronchitis, burning sensation, cough, diabetes, erysipelas, leprosy, skin diseases, ulcers.

24. *Albizia procera* Benth. (Mimosaceae)

Vern.: Safed Siris.

Distribution: In mixed forests & roadsides. Balaghat, Damoh, Gwalior, Hosahangabad, Jabalpur, Mandla, Seoni, Sidhi.

Botany: Trees, with grey-white bark, 2-6 pairs of pinnae and 6-15 pairs of ovate-oblong leaflets. Flowers pale-white in panicles. Pods brown.

Flowering & Fruiting: August-May.

Medicinal Parts: Leaves.

Chemical Constituents: Seed contain Saponins.

Seed oil contain Fatty acid.

Leaves contain Tannins, Saponins, Hentriacontane, Hexacosanol, α-Spinasterol, Carotenoid, Protein, Oxalate, Nitrate.

Gum contain D-galactose, L-arobinose, L-rhamnose.

Heartwood contain 4-Isoflavones; Biochanin-A, Formononetin, Genistein daidzein.

Medicinal Properties: Leaves – Insecticide.

Medicinal Uses: Leaves – Ulcers.

25. *Allium cepa* L. (Liliaceae)

Vern.: Pyaz.

Distribution: Cultivated throughout the State.

Botany: An annual herb, with bulbous stem, fistular, linear leaves and white flowers in umbels.

Flowering & Fruiting: December-March.

Medicinal Parts: Leaves, Bulb..

Chemical constituents: Bulb contains Glutamyl, Arginin, Cystein, Quercetin, Kaempferol, Methylcystein, Sulphoxide, Propylecystein, Saponins.

Medicinal Properties: Aphrodisiac, diuretic, expectorant, stimulant.

Medicinal Uses: Anti-tumour, bronchitis, dysentery, gum swelling, Jaundice, night blindness, piles.

26. *Allium sativum* L. (Liliaceae)

Vern.: Lahsun.

Distribution: Cultivated throughout the State.

Botany: Bulbous herbs, each bulb with several fleshy bulblets. Leaves linear. Flowers white, in umbels.

Flowering & Fruiting: December-March.

Medicinal Parts: Leaves, Bulb.

Chemical constituents: Bulb contains A volatile oil which contains Allyl disulphide, Allyl-propyl disulphide and Poly sulphide; Allicin (antibacterial): Allicetion-I and Allicetion-II (Antibiotics).

Medicinal Properties: Antibacterial, antiseptic, aphrodisiac, bone healing, digestive, diphoretic, diuretic, expectorant, pungent, rejuvinator, stimulant, stomachic, tonic.

Medicinal Uses: Asthma, bronchitis, chronic cough, ear trouble, epilepsy, flatulence, malaria fever, skin diseases, tuberculosis, ulcer.

Ayurvedic preperations: *Lasunadivati, Lasunadi ghirit, Rasonpinda, Rasonashtak, Majunseer, Rasonasura, Rasona vati, Rasona siddha taila.*

27. *Aloe vera* L. (Liliaceae)

Syn.: *A. barbadensis* Mill.; *A. perfoliata* L. var. *veral*

Vern.: Gwarpatha.

Distribution: Cultivated throughout the State, sometimes wild in drier localities.

Botany: Perennial herbs, with fleshy mucilaginous leaves in rosette having marginal spines. Flowers yellow, born in racemes on long scapes.

Flowering & Fruiting: November-March.

Medicinal Parts: Whole plant – dried juice & pulp of leaves.

Chemical constituents: Leaves contain Carbohydrates, Tannins, Proteins, Chromones (Abesin & Aloesone), Malic acid, Citric acid, Tartaric acid.

Jelly from leaves contain Acetylated glucomannans, Iron, Calcium, Potassium & Magnesium.

Medicinal Properties: Leaf pulp – Analgesic, anthelmintic, anti – inflammatory, carminative, digestive, purgative, stimulant, stomachic, tonic.

Medicinal Uses: Leaf pulp – Cold, colic, constipation, cough, fever, menstrual disorders, piles, rectal fissures, rheumatism.

Ayurvedic preperations: *Kumariasava, Kumarikavati, Kumaripak, Rajparvatni vati, Habbshabyar, Habb-ayariz, Habb sibra, Habbantkar, Suful dama.*

28. *Alstonia scholaris* R. Br. (Apocynaceae)

Syn.: *A. cuneata* Wall; *Echites scholaris* L.

Vern.: Saptparni.

Distribution: Planted. Indore.

Botany: A large tree with whorled branches and leaves with bitter milky juice, white flowers.

Flowering & Fruiting: December-June.

Medicinal Parts: Leaves, Bark, Wood.

Chemical constituents: Leaves contain Picralinol, Akuammigine, Tubotaiwine, Akuammicine.

Flowers contain Picrinine, Strictamine, Tetrahydro alstonic, Hexacosane, Lupeol, Beta-Amyrin, Palmitic acid, Urolic acid.

Stem bark contain Echitamine, Venoterpine-Glucoside.

Medicinal Properties: Antileprotic, astringent, blood purifier, liver tonic, sedative, stomachic.

Medicinal Uses: Dropsy, dysentery, gastro-intestinal troubles, malaria, ulcers.

Ayurvedic preperations: *Saptaparnasatvadi vati, Saptchhadadi kwath, Malariasambhar vati, Saptacchadadi taila, Saptaparna ghanasara.*

29. *Alternanthera sessilis* (L.) R. Br. ex. DC. (Amaranthaceae)

Syn.: *A. triandra* Lam.; *A. denticulata* R. Br.; *A. nodiflora* R. Br.

Vern.: Guroo.

Distribution: Common in aquatic and semi-aquatic situations. Balaghat, Betul, Chhatarpur, Damoh, E. Nimar, Dhar, Guna, Gwalior, Hosahangabad, Indore, Jabalpur, Mandla, Panna, Rewa, Sagar, Satna, Seoni,, Shivpuri, Sidhi, Tikamgarh.

Botany: A prostrate or spreading, polymorphic herbs; stem grooved, greenish-purple, with white hairy appendages, much branched, often rooting at the nodes; leaves glabrous, elliptic, linear-oblong, cuneate at the base, acute; flowers in axillary, small, white or pink-tinged, spicate-heads; seeds minute.

Flowering & Fruiting: Throughout the year.

Medicinal Parts: Whole plant.

Plant contains Iron, protein, hydrocarbons & sterols.

Medicinal Properties: Dried plant – Abortifacient, febrifuge.

Medicinal Uses: Dried plant – Blood disorders, leprosy.

Root – Wound.

30. *Amaranthus spinosus* L. (Amaranthaceae)

Vern.: Kantewali chauli.

Distribution: Common in wastelands and disturbed area along roadsides. Balaghat, Chhatarpur, Damoh, Indore, Jabalpur, Mandla, Panna, Rewa, Satna, Shahdol, Sidhi, Tikamgarh, W. Nimar.

Botany: Monoecious herb, with ovate-rhomboid or elliptic leaves having axillary spines. Flowers in axillary clusters, passing to spikes. Fruits ovoid.

Flowering & Fruiting: Throughout the year.

Medicinal Parts: Root and leaves.

Chemical constituents: Plant contains Carbohydrates, Proteins, Minerals, Iron, Vitamin – A and Calcium (Leaves).

Medicinal Properties: Alexiteric, antipyretic, appetiser, cooling, digestive, diuretic, laxative, stomachic.

Medicinal Uses: Biliousness, blood diseases, burning sensation, bronchitis, hallucination, leprosy, leucorrhoea, piles.

31. *Amaranthus tricolor* L. (Amaranthaceae)

Syn.: *A. tristis* L.; *A. gangeticus* L.; *A. melancholius* L.; *A. polygamus* L.

Distribution: Common as weed in wastelands & Cultivated fields, also cultivated. Balaghat, Chhatarpur, Damoh, Gwalior, Mandla, Satna, Shahdol, Sidhi, W. Nimar.

Botany: Ascending herbs; Stems angular; flowers densely Clustered in interrupted spikes.

Flowering & Fruiting: August-October.

Medicinal parts: Whole plant.

Chemical constituents: *Leaves* – Amino acids, Calcium, Magnesium, Mangnese, Phosphorus, Iron, Amaranthine, Cholesterol, Campesterol, Stigmasterol, Fucosterol, Glucose, Fructose, Isoamaranthine.

Medicinal properties: Astringent, Appetizer, Digestive, Diuretic.

Medicinal uses: Dysentery, Haemoptysis.

32 *Ammania baccifera* L. (Lythraceae)

Syn.: *A. salicifolia* Hiern.; *A. vesicatoria* Roxb.

Distribution: Balaghat, Damoh, Gwalior, Hosahangabad, Indore, Jabalpur, Mandla, Morena, Narsimhapur, Rewa, Sagar, Satna, Seoni, Sidhi.

Botany: Erect or ascending herbs, with oblong or elliptic leaves and sessile flowers lacking petals and style. Capsules globose, reddish.

Flowering & Fruiting: November-April.

Medicinal Parts: Whole plant.

Chemical constituents: Hentriaconite, Dotriacontanol, Betulinic acid, Lupeol, Ellagic acid, Quercetin, Lawsone.

Medicinal Properties: Plant – Acrid, aphrodisiac, appetizer, bitter, cooling, diuretic, laxative, rubifacient, stomachic.

Medicinal Uses: Plant – Burning sensation, colic, dyspepsia, fever. Flatulence, rheumatism, seminal weakness, strangury.

33. *Amorphophallus campanulatus* (Roxb.) Blume ex Decne (Araceae)

Vern.: Jangali suran.

Distribution: In shady places. Indore.

Botany: A perennial herb. Tubers depressed-globose, bulbiferous, dark brown. Leaves very large, arising after the flowers, long with pale blotches, stout; leaflets obovate; flowers in long spadix.

Flowering & Fruiting: July-October.

Medicinal Parts: Corm.

Chemical constituents: Corm contains Proteins, fats, Carbohydrate, Minerals, Calcium Oxalate, Phosphorous, Iron, vitamin A & B.

Medicinal Properties: Corm – Acrid, anthelmintic, appetiser, carminative, digestive, expectorant, pungent, stimulant, stomachic.

Medicinal Uses: Corm – Asthma, colic, cough, dysentery flatulence, gastric trouble, piles, rheumatism, vomiting, worms.

34. *Ampelocissus latifolia* (Roxb.) Planch. (Vitaceae)

Syn.: *Vitis latifolia* Roxb.

Vern.: Panibel.

Distribution: Balaghat, Damoh, Hosahangabad, Mandla, Rewa, Satna, Seoni, Shivpuri.

Botany: Large climbers, with hollow, striate branches. Leaves orbicular-cordate, lobed, serrate. Flowers reddish-brown, in pyramidal thyrsoid cymes.berries globose, black, 2-seeded.

Flowering & Fruiting: August-November.

Medicinal Parts: Root.

Chemical Constituents: Root contains Saponins

Medicinal Properties: Root – Astringent.

Medicinal Uses: Root – Chronic, Dysentery, Wounds.

Leaf – Dental problem, Indolent ulcers, Wounds.

35. *Anacardium occidentale* L. (Anacardiaceae)

Vern.: Kaju.

Distribution: Cultivated for its edible cotyledons. Bhopal, Narsimhapur, Seoni.

Botany: A small tree; leaves coriaceous, obovate-oblong or elliptic, rounded at the apex, glabrous; flowers white, tuning red with age, in terminal panicles; nuts reniform.

Flowering & Fruiting: January– March.

Medicinal Parts: Root, bark, fruits.

Chemical constituents: Nut contain Polysaccharides (Galactose), Galact, Uronic acid.

Leaves contain P–Hydroxybenjoc acid, Protocatechuic acid, Gentistic acid, Gallic acid, Glucosides, Rhamnosides, Arabinosides, Xylosides, Kaempferol, Quercitol.

Medicinal Properties: Root – Puragtive.

Fruit – Antidiarrhoeal, emollient, nutritive, rubefacient.

Stem bark – Alterative, astringent.

Medicinal Uses: Fruit – Dropsy, dyspepesia, fever, kidney trouble, piles.

Bark – Leprosy, ulcer.

36. *Andrographis paniculata* (Burm. f.) Wall. ex Nees (Acanthaceae)

Syn.: *A. subspathulata* Clarke; *Justicia Paniculata* Burm.

Vern.: Kalmegh.

Distribution: Common in wastelands. Balaghat, Bhopal, Chhatarpur, Damoh, Hosahangabad, Narsimhapur, Sidhi.

Botany: Erect, glabrous, annual. Branches 4-angled. Leaves slightly undulate; base tapering. Flowers small, solitary, whitish, spotted with rose-purple. Corolla 2-lipped. Capsules linear-oblong.

Flowering & Fruiting: September-April.

Medicinal Parts: Whole plant, leaves, roots.

Chemical constituents: Plant contains Kalmeghin, Andrographolid, Tannic acid, Sodium and Potassium.

Medicinal Properties: Antiperiodic, digestive, Hepatic antileprotic, nutritive, stimulant, stomachic.

Medicinal Uses: Anorexia, blood purifier, hepatomegally, malaria, oedema, skin diseases.

Ayurvedic preperations: *Visamjwarhar kshar, Sudarshan churna, Kalmegh-navayas churna, Kalmegh kwath, kalmegh svarasa, Panchanga churna, Switradilepa.*

37. *Anethum graveolens* L. (Apiaceae)

Syn.: *Peucedanum graveolens* (L.) Hiren.; *P. sowa* Roxb.; *Anethum sowa* Roxb. ex Flem.

Vern.: Sowa.

Distribution: Chhatarpur.

Botany: A glabrous aromatic annual herb; leaves pinnately compound; flowers yellow, in compound umbels.

Flowering & Fruiting: January-March.

Medicinal Parts: Fruits.

Chemical constituents: Fruits contain Vicenin, Phellandrene, Dillapiol, Carvone, Dihydro carvone, D-limonene, A new xanthone glycoside-dillanoside.

Seed oil contain Essential oil, Tripetroselinin, Petroselinicdiolein, Dipetroselinicolein.

Medicinal Properties: Fruits – Acrid, alexeteric, anodyne, anthelmintic, antidysenteric, anti-inflammatory, antispasmodic, bitter, carminative, digestive, diuretic, expectorant, febrifuge, stomachic, sudorific, thermogenic.

38. *Anisomeles indica* (L.) Kuntze. (Lamiaceae)

Syn.: *A. ovata* R. Br.

Distribution: As forest undergrowth, on hill slopes and roadsides. Balaghat, Chhindwara, Damoh, Gwalior, Hosahangabad, Indore, Mandla, Panna, Rewa, Sagar, Seoni, Sidhi.

Botany: Suffruticose herbs, with ovate, acute or obtuse, crenate-serrate, softe, softly pubescent leaves. Flowers bluish-purple. Nutlets broadly ovoid, with rounded ends, polished. *Flowering & Fruiting:* August-March.

Medicinal Parts: Whole plant.

Chemical Constituents: The plant contains Essential oil.

Medicinal Properties-The plant has astringent, carminative, tonic properties.

Medicinal Uses: Essential oil from plant – Uterine affections.

39. *Annona reticulata* L. (Annonaceae)

Vern.: Ramphal.

Distribution: Planted in Gardens. Bhopal, Mandla.

Botany: Small trees; flowers yellow-green; ripe fruits round yellowish brown. *Flowering & Fruiting:* September-May.

Medicinal Parts: Bark, fruits, seeds, leaves.

Chemical constituents: Root bark contains – Alkaloids: Oxoushinsunine, Iiriodenine, anonine, norushinsunine, michelabine, reticulin.

Fruits – Protein, fat, minerals: Calcium, phosphorus, iron, riboflavin; vitamin-C & niacin. Carotene.

Medicinal Properties: Root – Insecticidal, purgative.

Seed – Stringent, insecticidal.

Leaves – Anthelmintic, insecticidal.

Stem – Insecticidal.

Bark – Astringent, vermifuge.

Fruit – Anthelmintic, antidysentric, sweet.

Medicinal Uses: Leaves – Inflammatory tumors.

Bark – Dysentery, tonic.

Fruit – Diarrhoea, dysentery.

40. *Annona squamosa* L. (Annonaceae)

Vern.: Seetaphaal.

Distribution: Cultivated in gardens for its edible fruits as well as an escape along forest margins. dry deciduous forests and roadsides. Balaghat, Damoh, Dhar, Hosahangabad, Indore, Mandla, Rewa, Seoni, Tikamgarh.

Botany: Evergreen shrubs or small trees with oblong-lanceolate leaves and yellowish green flowers. Fruits globose or ovoid with projecting areoles.

Flowering & Fruiting: July – November.

Medicinal Parts: Roots, Leaves, fruits & Seeds.

Chemical constituents: Leaves contain Anonaine, Roemerine, Corydine, Narcorydine, Norlaureline, Isocorydine, Noricocorydine, Glucine, Friedeline, Alkaloids, Essential oils, Volatile oil, Carvone, Michelabine, Liriodedine, Diazepine, Squamolone, Beta–Caryophylline, β-sitosterol, Camphor, Barmeol.

Medicinal Properties: Root – Purgative.

Bark – Astringent.

Leaves – Anthelmintic, antispasmodic, stimulant, sudorific.

Fruits – Antiscorbutic, astringent, Cooling, febrifuge, pectoral.

Seeds – Abortifacient, emetic.

Medicinal Uses: Bark – Diarrhoea.

Leaves – Boils, swelling, ulcers.

Fruits – Diarrhoea, dysentery, dyspepsia, vertigo, vomiting.

41. *Anogeissus latifolia* (Roxb. ex DC.) Wall. ex Guill. & Perr. (Combretaceae)

Vern.: Dhawra.

Distribution: In dry deciduous forests. Balaghat, Bhopal, Damoh, E. Nimar, Guna, Hosahangabad, Indore, Jabalpur, Mandla, Narsimhapur, Panna, Raisen, Rewa, Sagar, Satna, Seoni, Shahdol, Shivpuri, Sidhi, Tikamgarh.

Botany: Tall, deciduous trees, with smooth, white-grey bark. Leaves silky tomentose when young. Flowers 2-5 in a head, yellow to pinkish – yellow. Fruits orbicular, glabrous, with entire wings.

Flowering & Fruiting: October – June.

Medicinal Parts: Root, bark, fruits.

Chemical constituents: Seed contain Protein.

Leaves contain Phenol, Carboxylic acid, Saponin, Amino acids, Sugars.

Stem bark contain Shikimic acid, quinic acid, fructose, alanine, phenylalanine, tannins. Saponins.

Sap wood contain Quinic acid, glucose, phenylalanine, ellagic acid.

Heart wood contain Quercetin, myricetin, glucose, ellagic acid, gallic acid, phenylalanine.

Medicinal Properties: Root – Acrid, pungent, stomachic.

Bark – Acrid, appetiser, cooling, pungent, sweet.

Medicinal Uses: Bark – Anaemia, erysepelas, piles, skin diseases, urinary discharges.

42. *Anthocephalus cadamba* (Roxb.) Miq. (Rubiaceae)

Syn.: *A. chinensis* Lam; *A. indicus* A. Rich.

Vern.: Kadam.

Distribution: Mandla, Rewa, Satna.

Botany: A large tree with spreading branches; leaves elliptic-oblong, glabrous above; flowers sweet scented, fruits fleshy.

Flowering & Fruiting: June – September.

Medicinal Parts: Bark, leaves, fruits.

Chemical constituents: Leaves contain Alkaloid (Kadambine & 3 – œ – di – hydrocadambine); Glycosides alkaloids (Iso – dihydrocadambine & β – Sitosterol).

Root & stem bark contain Quinovic acid, Pentacyclic triterpenic acid, cadambagenic, quinioric acid, saponin.

Flower contain Essential oils.

Medicinal Properties: Acrid, anodyne, astringent, anti – inflammatory, bitter, carminative, constipation cooling, digestive, diuretic, expectorant, febrifuge.

Medicinal Uses: Bark – Cough, debility, diarrhoea, dyspepsia, fever, inflammation, ulcers, vomiting, wound.

Leaves – Ulcers, wounds.

Fruits – Fever.

Ayurvedic preperations: *Kadamba tvak churna, Kadamba phala rasa.*

43 ***Arachis hypogaea*** **L. (Fabaceae)**

Syn.: *A. asiatica* Lour.

Vern.: Mung-phali.

Distribution: Cultivated for seeds which yield edible oil. Bhopal, Chhindwada, Indore, Shivpuri, Sidhi.

Botany: Semi erect much branched slightly hairy annual herbs.

Flowering & Fruiting: August-November.

Medicinal Parts: Seed, oil.

Chemical constituents: Seed contains α-Resorcylic acid, P-hydroxybenzoic acid, Cis & Trans – P – Coumaric acid, Phloretic acid, Vanillic acid.

Medicinal Properties: Seeds – Aphrodisiac, constipating, sweet.

Oil – Emollient, purgative, sweet.

Medicinal Uses: Seeds – Diarrhoea, general debility.

44. ***Areca catechu*** **L. (Arecaceae)**

Vern.: Supari.

Distribution: Cultivated. Bhopal (Oommachan, 1977).

Botany: A slender tall palm tree, 40-60 feet high; flowers in large spadix.

Flowering & Fruiting: Throughout the year.

Medicinal Parts: Roots, Leaves, Fruits.

Chemical constituents: Nuts contain Alkaloids: Arecoline, Arecaidine, Guvacine, Isoguvacine, Leucocyanidin.

Medicinal Properties: Nuts – Anthelmintic, aphrodisiac, astringent, cooling, digestive, diuretic.

Medicinal Uses: Nuts – General debility, urinary disorders.

45. *Aristolochia bracteolata* Lam. (Aristolochiaceae)

Vern.: Kidamar.

Distribution: In wastelands. Bhopal, Shivpuri.

Botany: Perennial prostrate herbs, with purple flowers and capsule fruits. *Flowering* & *Fruiting:* September-March.

Medicinal Parts: Roots, Leaves.

Chemical constituents: Leaves contain Ceryl alcohol, β-Sitosterol, Aristolochic acid, KCl.

Roots contain Aristolochic acid, KCl, KNO_3.

Medicinal Properties: Roots & Leaves – Acrid, anthelmintic, anti-inflammatory, antiperiodic, appetizer, bitter, cathartic, sudorific, thermogenic, vulnerary.

Medicinal Uses: Roots & Leaves – Amenorrhoea, boils, colic, constipation, dysmenorrhoea, dyspepsia, eczema, inflammations, intsestinal worms, gonorrhoea, skin diseases, syphilis.

46. *Aristolochia indica* L. (Aristolochiaceae)

Syn.: *A. lanceolata* Wight.

Vern.: Isari mul.

Distribution: In mixed forests. Chhatarpur.

Botany: Twining, woody herbs, with elliptic-ovate or obovate – panduriform leaves. Flowers greenish-white, in axillary racemes. Capsules oblong-globose; seeds winged.

Flowering & Fruiting: July-February.

Medicinal Parts: Whole plant.

Chemical constituents: Plant contains Three bases of Arisotolochine, Nitrogenous acids (Aristinic acid, Aristidinic acid & Aristolic acid); Volatile oil.

Medicinal Properties: Analgesic, anthelmintic, anti-inflammatory, antipyretic, cardiotonic, purgative, stimulant, stomachic, tonic.

Medicinal Uses: Amenorrhoea, erysepelas, fever, flatulence, rheumatism, scorpion sting, snakebite, spider bite, strangury, ulcer, worms.

47. *Argemone mexicana* L. (Papaveraceae)

Vern.: Satyanashi.

Distribution: In waste places. Common throughout the State.

Botany: Prickly herbs, with yellow sap and pinnatifid and spiny leaves. Flowers yellow, solitary terminal. Capsule prickly.

Flowering & Fruiting: February-June.

Medicinal Parts: Root, yellow latex.

Chemical constituents: Root contains Berbetine, Protopine nitrate, Betasterol, fatty acids, ceryl alcohol, Allocryptopine.

Flowers contain Tannin, Resin, Argemone oil.

Leaves contain Glucosides, Kampferin, Anthraquinone, essential oil & Chrysophanic acid.

Seeds contain tasteless oil of dark brown colour, Carbohydrates, Albumin, Phosphate, sulphate.

Medicinal Properties: Root – Alterative, anodyne, Diuretic.

Seed – Demulcent, emetic, expectorant, laxative.

Medicinal Uses: Root – Skin diseases.

Seed – Diarrhoea, dysentery.

Ayurvedic preperations: *Svarnaksiri bija churna & Taila.*

48. *Argyreia speciosa* Sweet. (Convolvulaceae)

Syn.: *A. nervosa* (Burm. f.) Bojer.

Vern.: Samander ka pat

Distribution: Rare. Hoshangabad.

Botany: White tomentose twiners, with ovate-cordate leaves, tomentose beneath. Flowers rose-purple and white.

Flowering & Fruiting: April-September.

Medicinal Parts: Whole plant, root, leaves, seeds.

Chemical constituents: Plant contains Arginine, Lycine, Valine, Behenic acid, Linolic acid, linolenic acid, Oleic acid, Argoclavine, Clanoclavine–I & II, Elymoclavine, Ergine, Isoergine, Ergometrine, fastuclavine, Isosetoclavin, setoclavine, Lysergine, Lysergol, Lysergic acid, Caffeic acid, Anthocyanin, Cumarine, Flavonoids, Quercetine, Kaempferol, β – Setosterol.

Medicinal Properties: Leaf – Emollient.

Root – Alterative, bitter, diuretic.

Medicinal Uses: Leaf – Eczema, itching, ringworm, skin diseases.

Root – Gonorrhoea, rheumatism, syphilis.

Ayurvedic preperations: *Vrddhadarukasama churna, Vrddhadaruka rasayana, Laksmivilasa guti.*

49. *Artemisia nilagirica* (Clarke) Pamp. (Asteraceae)

Syn.: *A. vulgaris* auct. (non L.)

Distribution: Balaghat, Hosahangabad. Common in mixed forests.

Botany: A tall aromatic perennial shrub; lower leaves deeply pinnatified, uppermost smaller, trifid or entire, lanceolate; flowers in subglobose heads; fruits achenes.

Flowering & Fruiting: Ocober-September.

Medicinal Parts: Leaves, Flowering tops.

Chemical constituents: Essential oil contain P-cymene, L-linalool, β-thujone, Azulene, Thujyl alcohol and an unidentified Hydrocarbon.

Artemisia Ketone, a Pentacyclic Triterpene, Thujone, a-amyrin & its acetate, Ferneol, Stigmasterol and β-sitosterol.

Medicinal Properties: Leaves & Flowering tops – Acrid, alexeteric, anodyne, anthelmintic, anti-inflammatory, aphrodisiac, appetiser, aromatic, astringent, bitter, depurative, digestive, diuretic, febrifuge, stomachic, thermogenic.

Medicinal Uses: Leaves & Flowering tops – Amenorrhoea, asthma, bronchitis, colic, cough, dysmenorrhoea, dyspepsia, fever, flatulence, hysteria, inflammations, intestinal worms, leprosy, pruritus, skin diseases, strangury.

50. *Artocarpus heterophyllus* Lam. (Moraceae)

Vern.: Kathal.

Syn.: *A. integrifolia* auct. non Linn.

Distribution: Cultivated throughout the State. Chhatarpur.

Botany: Tree with dense, cariaceous elliptic leaves. Stipules glabrous. Male receptacle cylindric and female receptacle tuberculed. The receptacles are lateral on trunk and branches, on special branchlets. The inflorescence is enclosed in 2 large, lathery yellowish stipules belonging to the leaf.

Flowering & Fruiting: November-June.

Medicinal Parts: Root, leaves, fruits, seeds, wood, latex.

Chemical constituents: Leaves contain Cycloartenone, cycloartenol, B-sitosterol, tannin.

Fruit contains B-carotene, amino acids, aurantiamide acetate, vitamin – A & C; iron phosphorus.

Bark contains Betulinic acid, cycloheterophyllin, iso cycloheterophyllin, cycloartenyl acetate, cycloartenol, cycloartenone.

Heartwood contains Morin, dihydromorin, cyanonaclurin, artocarpin, isoartocarpin, cyclo artocarpin, artocarpesin, cycloartocarpesin, artocarpetin, artocarpanone, Sucrose, ellagic acid.

Latex contain Cycloartenon, cycloartenol, B-sitosterol, butyrospermol, leucine, isoleucine, tyrosine, valine, cycloartenol.

Medicinal Properties: Unripe fruit – Acrid, astringent, carminative, tonic.

Ripe fruit – Aphrodisiac, cooling, latex, sweet, tonic.

Seeds – Aphrodisiac, constipating, diuretic, sweet.

Medicinal Uses: Unripe fruit – Dyspepsia, general debility.

Ripe fruit – Biliousness, burning sensation, constipation, seminal debility, ulcers, tonic.

Seeds – General tonic, tuberculosis.

Latex – Abscesses, swellings.

51. *Arundo donax* L. (Poaceae)

Distribution: Gwalior, Indore, Mandla, Rewa.

Botany: Perennial plant, with light greenish yellow flowers and caryopsis fruits.

Flowering & Fruiting: August-November.

Medicinal Parts: Rhizome.

Chemical Constituents: Rhizome contains – Indole bases, Dehydro-bufotenine, Bufotenidine, Gramine (Donaxine).

Medicinal Properties: Rhizome – Aphrodisiac, Cooling, Diuretic, Emollient, Galactagogue, Hypotensive, Sudorific, Sweet.

Medicinal Uses: Rhizome – Blood diseases, Dysuria, Erysipelas, Eye diseases, Leucorrhoea, Piles.

52. *Asclepias curassavica* L. (Asclepiadaceae)

Distribution: Planted. Mandla.

Botany: An erect, simple or much-branched perennial undershrub, with a woody base. Flowers orange-yellow in many flowered umbels. Fruits follicles.

Flowering & Fruiting: Throughout the year.

Medicinal Parts: Whole plant, Root, Leaves, Flowers.

Chemical constituents: Leaves contain β-sitosterol, β-amyrin, myristic acid, palmitic acid, stearic acid, oleic acid, linoleic acid, arachidic acid, arachidonic acid, glucose, fructose, acidic mucilage, asclepine glycoside.

Medicinal Properties: Plant – Abortifacient, anthlemintic, antipyretic, cardio tonic, emetic, expectorant, hypertensive, purgative, styptic, sudorific.

Medicinal Uses: Plant – Gonorrhoea, haemorrhagic diseases, piles.

53. *Asparagus racemosus* Willd. (Liliaceae)

Syn.: *A. javanica* Kunth.; *A. sarmentosus* Graham.

Vern.: Shataveri.

Distribution: Throughout the State.

Botany: Scandent, spinous undershrubs, with faciculated tuberous roots. Leaves absent, branches modified into linear, leaf-like cladodes. Flowers in racemes. Fruits globose.

Flowering & Fruiting: August-January.

Medicinal Parts: Fleshy roots.

Chemical constituents: Fleshy roots contain Sugars, mucilage & saponins.

Medicinal Properties: Roots – Alterative, aphrodisiac, appetiser, astringent, Bitter, cooling, indigestion, stomachic, sweet, tonic.

Medicinal Uses: Root – Biliousness, diseases of the blood & the eye, dysentery, epilepsy, inflammation, leprosy, night blindness, throat complaint, tuberculosis, tumour.

Ayurvedic preperations: *Shatavari ghrita, Phalghriti, Visnu taila, Shatmulyadi louh, Shatavari panak, Narayan taila, Dhatupaushtik churna, Shatavari churna, Supari pak, Lakshmivilas ras, Basant kusumakar ras, Phalkalyan ghirit, Saraswatrisht, Geriforte, Halwa-i-supari pak, Halwa-i-ghaikwar, Majun zanjibil, Majun shir bargadh wali, Majun musli pak, Shatavari kalp, Shatavari yoga.*

54. *Averrhoa carambola* L. (Oxalidaceae)

Distribution: Cultivated for its edible fruits.

Botany: A medium sized evergreen tree; leaves compound; flowers pinikish to purplish in short racemes; fruits berries; seeds arillate.

Flowering & Fruiting: March-August.

Medicinal Parts: Leaves, Fruits.

Chemical constituents: Fruit contains Proteins, Amino acid compounds, Vitamin-C, Oxalic acid.

Flowers contain Rutin, Quercetin-3-glucoside.

Leaves contain Oxalic acid, potassium oxalate, fluorine, glucose, sucrose, fructose, vitamins.

Medicinal Properties: Leaves – Anthelmintic, antipruritic, antipyretic.

Fruits – Antipyretic, constipating, febrifuge, sour, sweet, thermogenic.

Medicinal Uses: Leaves – Fever, intestinal worms, pruritus, scabies.

Fruits – Diarrhoea, fever, general debility, haemorrhoids, scabies, vomiting.

55. *Azadirachta indica* A. Juss. (Meliaceae)

Syn.: *Melia azadirachta* L.

Vern.: Neem.

Distribution: Almost throughout the State. Commonly Planted.

Botany: Evergreen trees, with serrate or crenate, 7-17 leaflets and white flowers in panicles. Drupes ovoid, yellow when ripe.

Flowering & Fruiting: March – May.

Medicinal Parts: Whole plant, root bark, stem bark, leaves, flowers, fruits.

Chemical constituents: Plant contains Margocin, Nimbadin, Sulphur, Nimban, Nimbosterol, Tannin, essential oil, Resin, Glycosides, Fatty acids, gum, free amino acids.

Leaves contain Quercetin, β – Sitosterol (Nimbasterol).

Medicinal Properties: Stem bark – Anthelmintic, astringent, Bitter, demulcent, purgative, tonic.

Leaves – Anthelmintic, anti – inflammatory, antileprotic, stimulant.

Medicinal Uses: Stem bark – Burning sensation, fever, itching, skin diseases, Ulcer, urinary disorders.

Leaves – Diseases of the eye, fever, inflammation, skin diseases, ulcer, urinary disorders.

Ayurvedic preperations: *Nimbadi kwath, Nimbadi churna, Nimbarishta, Nimbaridrakhand.*

56. *Bacopa monnieri* (L.) Penn. (Scrophulariaceae)

Syn.: *Herpestis monnieri* (L.) H. B. & K.; *Moniera cuneifolia* Michx.; *Gratiola monniera* L.

Vern.: Brahmi.

Distribution: In marshy and swampy area. Chhatarpur, Damoh, Dhar, Gwalior, Mandla, Panna, Rewa, Shivpuri, Sidhi, W. Nimar.

Botany: A prostrate gregarious herb with spreading or ascending branches, rooting at the nodes. Leaves sessile, decussate, obovate-oblong or spathulate, succulent. Flowers axillary, solitary on slender pedicels. Corolla bluish –purplish; anthers bluish-purple. Capsules ovoid, glabrous.

Flowering & Fruiting: July-April.

Medicinal Parts: Whole plant.

Chemical constituents: Plant contains Alkaloids – Brahmine & Herpestine, Bacoside A & B (Saponins), Betulic acid, Hersaponin, Nicotine, Becogenin – A, Potassium, Sodium, Calcium, Iron, Oleic acid, Palmitic acid.

Medicinal Properties: Anti – inflammatory, blood purifier, diuretic, laxative, tonic.

Medicinal Uses: Blood diseases, cough, debility after heart attack, dyspepsia, epilepsy, fever, hoarseness of voice, insomnia, less memory, tension.

Ayurvedic preperations: *Brahmi ghirit, Brahmi prasa, Brahmi siddha taila, Sarasvatarishta.*

57. *Balanites aegyptiaca* (L.) Del. (Simaroubaceae)

Syn.: *B. rouxburghii* Planch.; *B. ingudi*; *Ximenia aegyptiaca* L.

Vern.: Hingot.

Distribution: Gwalior, Satna.

Botany: Spinous trees or shrubs, with 2-foliolate leaves. Flowers greenish-yellow, in axillary fascicled cymes. Fruits oblong, rind hard.

Flowering & Fruiting: October-August.

Medicinal Parts: Fruits, seeds.

Chemical constituents: Fruits & seeds contain Oil of yellow colour and tasteless.

Pulp & fruit contains Saponin.

Medicinal Properties: Flower – Bitter, heating, sweet smelling.

Fruits – Antidysenteric, alexipharmic, alterative, analgesic anthelmintic, bitter sharp digestive, expectorant, laxative, taste.

Leaf – Anthelmintic, bitter, laxative.

Medicinal Uses: Fruits – Cough, leprosy, skin diseases, ulcer, wound, worm infestation.

Leaf – Cough, throat infection.

58. *Baliospermum montanum* (Willd.) Muell-Arg. (Euphorbiaceae)

Syn.: *B. polyandrum* Wight.; *B. axillare* Blume.

Distribution: Common in shady places and along the bank of drains in mixed forests. Damoh, E. Nimar, Hosahangabad, Indore, Mandla, Narsimhapur, Panna, Rewa, Satna, Seoni, Sidhi.

Botany: An undershrub; with herbaceous branches from roots; leaves variable in shape & size; flowers small, greenish, monoecious in axillary racemes; fruits capsules.

Flowering & Fruiting: February-April.

Medicinal Parts: Roots, Leaves, Seeds.

Chemical constituents: Root contains 5 new phorbol esters belonging to diterpene hydrocarbotigliane skeleton viz. montain, balliospermin etc.

Leaves contain β-sitosterol, glucosides, hexacosanol.

Medicinal Properties: Roots – Acrid, anodyne, anthelmintic, anti-inflammatory, diaphoretic, digestive, diuretic, febrifuge, purgative, rubefacient, thermogenic.

Seeds – Purgative, rubefacient, stimulant.

Medicinal Uses: Roots – Anaemia, constipation, dropsy, fever, flatulence, haemorrhoids, jaundice, leprosy, leucoderma, skin diseases, strangury, wounds.

Seeds – Flatulence, inflammations.

59. Bambusa arundinacea (Retz.) Willd. (Bombusaceae)

Syn.: *B. bambos* Druce.

Distribution: Bhopal, Damoh, Hosahangabad.

Botany: Woody, thorny reed-bamboo, with stem-sheath hairy, cariaceous, striate. Leaves triangular-lanceolate, base rounded glandular. Spikelets pale, lanceolate, acute, glabrous. Caryopsis oblong, with persistent style-base, grooved. *Flowering & Fruiting:* Once in life.

Medicinal Parts: Roots, leaves, sprouts, grains.

Chemical constituents: Root contains Cyanogenetic glucoside.

Young shoot contains HCN, Benzoic acid, Reducing sugars & Waxes.

Medicinal Properties:

Roots – Astringent, cooling, depurative, diuretic, laxative, sweet.

Leaves – Astringent, constipating, cooling, febrifuge, sweet.

Sprouts – Acrid, anthelmintic, anti-inflammatory, bitter, carminative, digestive, diuretic, laxative.

Grains – Acrid, alexeteric, anthelmintic, aphrodisiac, sweet.

Medicinal Uses: Roots – Burning sensation, leprosy, ringworm, skin diseases.

Leaves – Diarrhoea, fever, haemorrhoids, lumbago, skin diseases, wounds.

Sprouts – Dyspepsia, flatulence, inflammations, intestinal worms, ulcers, wounds.

Grains – General debility, intestinal worms.

60. *Barleria prionitis* L. (Acanthaceae)

Vern.: Jhinti.

Distribution: Common as undergrowth in mixed forests. Bhopal, Chhatarpur, Damoh, Gwalior, Hosahangabad, Indore, Jabalpur, Morena, Panna, Raisen, Rewa, Sagar, Seoni, Shivpuri, Sidhi, Tikamgarh.

Botany: Armed undershrubs, with elliptic or ovate-elliptic leaves and yellow flowers in dense spikes. Calyx-lobes glandular. Capsules 4-seeded.

Flowering & Fruiting: September-April.

Medicinal Parts: Root, leaves.

Chemical constituents: Plant contains Beta – Sitosterol, Scutellarein – 7 Rhamocyl Glucoside, Iridoids, Barlerin and Ac – Barlerin.

Medicinal Properties: Acrid, Anti-inflammatory, Antiseptic, Appetiser, bitter, diaphoretic, diuretic, expectorant, heating.

Medicinal Uses: Leaf juice – Abdominal disorders, catarrha, fever.

Dried bark – Cough.

Decoction of plant – Cough, dropsy, Paralysis, toothache.

Paste of root – is applied on boils & glandular swelling.

61. *Barringtonia acutangula* (L.) Gaertn. (Borringtoniaceae)

Syn.: *B. spicata* Bl.

Distribution: Growing along the banks of rivers and streams.

Botany: A middle sized evergreen tree. Leaves elliptic, subsessile. Flowers in long pendulous racemes.

Flowering & Fruiting: July-November.

Medicinal Parts: Root, Leaves, Fruits.

Taste: Bitter, Acrid, Sweet.

Chemical constituents: Leaves contain Acutangulic acid, Barringtogenic acids, Stigmasterol, Tangulic acid, Oleanolic acid, β-amyrin, β-sitosterol, Stigmasterol, Sapogenin.

Fruits contain Triterpenoid sapogenins as Barringtogenol D, C& B, triterpenic acids as barrigenic acid, barrinic acid, barringtogenic acid.

Medicinal Properties:

Roots – Antipyretic, aperient, bitter, cooling, emetic, stimulant.

Leaves – Bitter, constipating.

Fruits – Acrid, alexeteric, anthelmintic, bitter, cooling, depurative, diuretic, emetic, expectorant, purgative.

Medicinal Uses: Root – Catarrh, constipation, fever.

Leaves – Diarrhoea, dysentery.

Fruits – Bronchitis, colic, cough, dysmenorrhoea, fever, intestinal worms, leprosy, lumbago, skin diseases, strangury, syphilis, ulcers, wounds.

62. *Basella alba* L. var. *rubra* (L.) Stewart (Basellaceae)

Syn.: *B. rubra* L.

Vrn.: Poi.

Distribution: Climber on bushes and hedges, often cultivated. Damoh, Sidhi.

Botany: A perennial climbing herb. Leaves broad ovate, cordate. Flowers pinkish or pale purple, sessile, in axillary, penduncled spkies. Flowers remain closed. Perianth fleshy, 5-lobed. Fruit enclosed in perianth.

Flowering & Fruiting: March-December.

Medicinal Parts: Stems, Leaves.

Chemical constituents: Plant contains Polysaccharides, organic acids, provitamin A, Carotenoids as α-carotene, β-carotene, violaxanthin, lutein, tunaxanthin & phytonene.

Medicinal Properties: Stem & Leaves – Aphrodisiac, appetizer, cooling, demulcent, diuretic, emollient, haemostatic, laxative, sedative, sweet.

Medicinal Uses: Stem & Leaves – Burning sensation, constipation, dysentery, gonorrhoea, leprosy, pruritus, strangury, ulcers, urticaria.

63. *Bauhinia purpurea* L. (Caesalpiniaceae)

Vern.: Kaliar.

Distribution: On roadsides and in mixed forests. Balaghat, Bhopal, Damoh, Gwalior, Indore, Jabalpur, Seoni, Sidhi.

Botany: Tree; branches stout, glaberescent; leaves roundish, divided half way down into 2 obtuse lobes; flowers fragrant in terminal and axillary few flowered racemes; pods subwoody, flat pointed at both ends, glabrous; seeds 12-16, oblong-ellipsoid, brown.

Flowering & Fruiting: September – November.

Medicinal Parts: Stem bark, flowers & leaves.

Chemical constituents: Flower contains Pelargonidin–3–glucoside, Pelargonidin – 3 – Triglucoside, Chalcon galactoside, Isoquercetin and Astragalin.

Medicinal Properties: Anthelmintic, astringent, carminative, cooling, laxative, rejuvenator, tonic.

Medicinal Uses: Flower – Blood dysentery, piles.

Stem bark – Diarrhoea.

Root bark – Cough, goitre, haemorroids, leucoderma.

64. *Bauhinia racemosa* Lam. (Caesalpiniaceae)

Syn.: *Piliostigma racemosa* (Lamk.) Bentham.

Distribution: In sal and mixed forests and roadsides. Balaghat, Chhatarpur, Chhindwara, Damoh, E. Nimar, Gwalior, Jabalpur, Hoshangabad, Mandla, Panna, Rewa, Sagar, Satna, Seoni, Shahdol, Shivpuri, Sidhi.

Botany: Deciduous trees.bark grey, rough. Leaves broader than longer. Flowers in terminal, leaf opposed racemes, creamy-white. Pods falcate, turgid. Seeds oval, black.

Medicinal Parts: Stem bark, leaves, flowers, seeds, fruits.

Chemical constituents: Stem bark contains Octacosane, B-amyrin, B-sitosterol.

Seeds – Protein, pentosan, mucilage.

Medicinal Properties: Root bark – Astringent.

Stem bark, Leaves – Antipyretic, alexipharmic, refrigerent.

Medicinal Uses: Root bark – Diarrhoea, dysentery, inflammation, skin diseases, ulcers.

Stem bark, Leaves – Biliousness, diseases of blood, fistula, headache, urinary discharges.

Flowers – Cough, haemorrhages, piles.

65. *Bauhinia variegata* L. (Caesalpiniaceae)

Vern.: Kachnar.

Distribution: In mixed forests. Balaghat, Bhopal, Chhindwada, W. Nimar, Indore, Mandla, Seoni, Shahdol, Shivpuri.

Botany: Medium-sized deciduous trees. Leaves orbicular, 2-lobed. Flowers white to pink. Pods flat, glabrous, brown.

Flowering & Fruiting: August– February.

Medicinal Parts: Stem bark & flowers.

Chemical constituents: Stem bark contains Sugar & Tannins.

Flowers contain Octacosanol, β – stiosterol, flavinoid.

Seeds & bark contain Myrcetol, glycoside, kaempferol, glucoside.

Medicinal Properties: Bark – Alterative, astringent, tonic.

Flowers – Acrid, astringent, cooling, sweet.

Medicinal Uses: Bark – Asthma, blood diseases, dysmenorrhoea, leprosy, leucoderma, stomach trouble, tuberculous glands, ulcers, wound.

Flowers – Blood diseases, bronchitis, headache, vaginal discharges.

Ayurvedic preperations: *Kanchanara guggulu, Kanchanara kwath, Kanchanara gutika.*

66. *Benincasa hispida.* (Thunb.) Cogn. (Cucurbitaceae)

Vern.: Gol kaddu.

Syn.: *B. cerifera* Savi; *cucurbita hispida* Thunb.

Distribution: Cultivated for fruits almost throughout the State.

Botany: A large climbing or trailing herb, with large, yellow, monoecious solitary flowers and large, fleshy, indehiscent fruits.

Flowering & Fruiting: June-October.

Medicinal Parts: Fruits, Seeds.

Chemical constituents: Fruit contains β-sitosterol, asparagine, luoeol, mannitol, proline, arginine, aspartic acid, glutamic acid, glutamine, hydroxyproline, Isoleucine, cysteine, L-leucine, n-triacontanol, glucose, rhamnose, vitamins.

Medicinal Properties:

Fruits – Aphrodisiac, antiperiodic, cooling, diuretic, laxative, styptic, sweet.

Seeds – Anthelmintic, cooling, sweet.

Medicinal Uses: Fruits – Asthma, cough, diabetes, epilepsy, fever, haemorrhages.

Seeds – Cough, fever, syphilis.

67. *Biophytum sensitivum* (L.) DC. (Oxalidaceae)

Syn.: *Oxalis sensitiva* L.

Vern.: Lajalu.

Distribution: Hosahangabad, Indore, Narsimhapur, Panna, Sagar, Satna, Seoni, Sidhi, Tikamgarh.

Botany: Erect unbranched annual herbs; leaves long, sensitive, leaflets opposite, sessile, oblong, terminal longest; flowers yellow, in terminal umbels; capsule ellipsoid; seeds ovoid.

Flowering & Fruiting: October – January.

Medicinal Parts: Whole plant, leaves, seeds.

Medicinal Properties: Anthelmintic, antiseptic, astringent, bitter, cooling, diuretic, stimulant, styptic, tonic.

Medicinal Uses: Asthma, chest complaint, cuts, fever, gonorrhoea, tuberculosis, wound.

68. *Blumea lacera* (Burm. f.) DC. (Asteraceae)

Syn.: *B. balsamifera* DC.; *B. densiflora* Hook. f.

Distribution: Common along the road sides, open fields and edges, on moist shaded places. Hosahangabad, Sagar, Seoni, Sidhi.

Botany: A herbaceous erect annual with a strong odour of turpentine; flowers yellow, numerous in heads, both terminal and axillary; fruits achenes, crowned with white pappus hairs.

Flowering & Fruiting: November-January.

Medicinal Parts: Whole plant.

Chemical constituents: The plant contains acetylenic compounds, Diester of Coniferol alcohol, Thiophenen derivatives, Campesterol, hentriacontane, Hentriaceontanol, α-amyrin and its acetate, Lupeol and its acetate, β-sitosterol.

Medicinal Properties: Plant – Acrid, anthelmintic, anti-inflammatory, astringent, bitter, digestive, diuretic, expectorant, febrifuge, styptic, thermogenic.

Medicinal Uses: Plant – Burning sensation, bronchitis, cough, fever, inflammation, intestinal worms, leucorrhoea, liver disorders, wounds.

69. *Boerhavia diffusa* L. (Nyctaginaceae)

Syn.: *B. repens* L.; *Boervia procumbens* Roxb.

Vern.: Punarnava.

Distribution: Common along roadsides, river banks and in wastelands. Chhatarpur, Damoh, E. Nimar, Indore, Seoni, Tikamgarh, W. Nimar.

Botany: Decumbent or diffuse, perennial, extensive herbs. Leaves unequal, ovate-oblong. Flowers rose or pink, in panicles or subcapitate umbels. Fruits clavate, 5-ribbed, covered with stalked glandular hairs.

Flowering & Fruiting: October-April.

Medicinal Parts: Fresh whole plant, root, leaves, flowers.

Chemical constituents: Plant contains Alkaloids – Punarnavan, Sterol, β – Sitosterol, œ – Sitosterol, Elentine, Stearic acid, Palmitic acid, Minerals, potassium nitrate, sodium & sulphate & chloride.

Medicinal Properties: Anti-bacterial, anti – inflammatory, antiviral, astringent, bitter, cooling, diaphoretic, laxative, stimulant, stomachic.

Medicinal Uses: Anaemia, asthma, biliousness, blood impurities, bronchitis, cough, epilepsy, eye trouble, heart diseases, inflammation, Jaundice.

Ayurvedic preperations: *Punarnava mandur, Punarnavakshar kwath, Punarnava ark, Punarnava ashtak, Punarnava asav, Punarnava churna, Calcury, Jawarish zaruni ambari ka naksha kalan, Punarnavambu, Punarnava guggula.*

70. *Bombax ceiba* L. (Bombacaceae)

Syn.: *B. malabaricum* DC.; *Salmalia malabarica* (DC.) Schott. & Endl.

Vern.: Semal.

Distribution: Damoh, Jabalpur, Sagar, Seoni, Sidhi.

Botany: Deciduous trees, with prickly trunk and digitately 5 to 7-foliolate leaves. Flowers crimson or yellow, crowded at the end of leafless branches. Capsules woody, 5-valved.

Flowering & Fruiting: February – May.

Medicinal Parts: Root, bark, flower, fruit, gum.

Chemical constituents: Seed contains Hexacosanol, tocopherol, terenes.

Bark contains Lupeol, B – sitosterol.

Root contains Triacontanol, B – sitosterol.

Flowers(Petals) contain Essential oil, Hentriacontane, Hentriacontanol, Quercetin, Kaempferol, B – sitosterol.

Medicinal Properties: Root – Astringent, cooling, demulcent, diuretic, sweet, tonic.

Bark – Acrid, astringent, demulcent, diuretic, tonic.

Flowers – Acrid, astringent, bitter, cooling.

Fruits – Alterative, aphrodisiac, cooling, digestive, diuretic, expectorant, stimulant, sweet, tonic.

Gum – Acrid, alterative, aphrodisiac, astringent, demulcent, tonic.

Medicinal Uses: Root – Biliousness, Inflammation.

Flower – Blood purifier, leucorrhoea.

Fruits – Leprosy.

Gum – blood diseases, diarrhoea, dysentery, leucorrhoea, monorrhagia.

Ayurvedic preperations: *Musli pak, Shalmali ghrit, Pushyanug churna, Brihadgangadhar churna, Musli simbhal, Mocarasa siddha ghrit.*

71. *Borassus flabellifer* L. (Arecaceae)

Syn.: *B. flabelliformis* Roxb.

Distribution: Cultivated. Dhar, Jabalpur.

Botany: A very tall, erect, magnificient dioecious palms; leaves 1-1.5m diam with 60-80 segments, folded along midrib; edges of petioles spinose; spadices very large; drupes large subglobose..

Medicinal Parts: Root, Leaves, Inflorescence, Fruits.

Chemical constituents: Stem juice contains Minerals like iron; protein, sucrose, glucose, thiamin, riboflavin, nicotinic acid, vitamin C.

Medicinal Properties: Roots – Cooling, diuretic, stimulant.

Fruits – Anthelmintic, aphrodisiac, cooling, depurative, laxative, sedative, stomachic, styptic, sweet.

Medicinal Uses: Root – Burning sensation, inflammations. Strangury.

Fruits – Burning sensation, colic, constipation, dyspepsia, flatulence, intestinal worms, leprosy, skin diseases.

72. *Boswellia serrata* Roxb. ex Colebr. (Burseraceae)

Syn.: *B. glabra* Roxb.

Vern.: Salai.

Distribution: Bhopal, Rewa, Seoni, Shahdol.

Botany: Deciduous tree, with bark peeling off in this flakes. Leaves imparipinnate, 7 to many-foliolate. Flowers white, bisexual. Drupes trigonous.

Flowering & Fruiting: February – April.

Medicinal Parts: Bark, gum-resin.

Chemical constituents: Bark contains Carbohydrates, Glucosides and β-sitosterol.

Gum resin contains Diterpine, alcohol, serratol, β-boswellic acid, essential oil.

Medicinal Properties: Bark – Antidysenteric, antipyretic, astringent, bitter, diaphoretic, diuretic, expectorant, stomachic, sweet.

Medicinal Uses: Bark – Asthma, dysentery, skin diseases, ulcers.

Gum-resin – Arthritis, asthma, bronchitis, cough, dysentery, fevers, jaundice, stomachache, syphilis.

73. *Brassica juncea* Czern. & Coss. (Brassicaceae)

Vern.: Rai.

Distribution: In Cultivated and escapes. Jabalpur, Satna, Sidhi.

Botany: A tall much-branched annual herbs. Basal leaves lyrate-pinnatipartile; middle ones obovate-oblong, pinnatilobed; upper ories lanceolate, entire-dentate. Racemes 20-40 flowered, up to 30 cm. long in fruit.

Flowering & Fruiting: October-March.

Medicinal Parts: Seeds, oil.

Chemical constituents: Seeds contain Sinigrin, Gluconapin, sinapine base, Isothyocynates, Essential oil, Flavonol glycosides.

Medicinal Properties: Seeds – Acrid, anodyne, anthelmintic, anti-inflammatory, bitter, carminative, digestive, sudorific, thermogenic.

Medicinal Uses: Seeds – Dyspepsia, flatulence, inflammations, intestinal worms, skin diseases.

74. *Brassica nigra* (L.) Koch. (Brassicaceae)

Syn.: *Sinapis nigra* L.; *S. erysimoides* Roxb..

Vern.: Banarasi Rai.

Distribution: In Cultivated. Indore.

Botany: A rigid, Branching annual upper leaves entire. Flowers bright yellow, in nacked racemes. Fruits siliqua. *Flowering & Fruiting:* December-March.

Medicinal Parts: Leaves, Seeds, Seed oil.

Chemical Constituents: Seed contains Glucoside Singirin, Sinapine & a Volatile Isothiocyanate and Essential oil.

Medicinal Properties: Seeds – Appetiser, Hot Sharp bitter taste.

Medicinal Uses: Seeds – Anthelmintic, Appetiser, Cough, Enlargement of spleen, Itching, Skin diseases, Tumours.

75. *Bridelia retusa* Spreng. (Euphorbiaceae)

Syn.: *B. airy-shawii* Li.; *B. spinosa* Willd.; *cluytia retusa* L.

Vern.: Ekdania.

Distribution: Common in mixed forests & along roadsides. Balaghat, Bhopal, Chhattarpur, Damoh, Dhar, Hosahangabad, Indore, Mandla, Panna, Satna, Seoni, Shahdol.

Botany: A shrub or a tree up to 18 m in height with 7 cm. long, strong conical spine.

Flowering & Fruiting: July-February.

Medicinal parts: Root, Stem, Bark.

Chemical constituents: *Fruit pulp* – β-Sitosterol, Gallic acid, Ellagic acids.

Seeds – Ellagic acid.

Bark: Triterpenoid Ketone, Tannin.

Medicinal properties: *Root & Bark* – Astringent, Antiviral, Hypoglycaemic & Hypotensive Properties.

76. *Buchanania lanzan* Spreng. (Anacardiaceae)

Vern.: Chironji.

Syn.: *B. latifolia* Roxb.

Distribution: Very common in dry deciduous forests. Balaghat, Bhopal, Hosahangabad, Indore, Jabalpur, Mandla, Rewa, Sagar, Satna, Seoni, Shivpuri, Tikamgarh.

Botany: Tall trees, with straight trunk, leaves elliptic-oblong. Flowers greenish-white, in pyramidal panicles. Drupes obiliquely lentiform, smooth, one-seeded.

Flowering & Fruiting: January-June.

Medicinal Parts: Roots, leaves, fruits.

Chemical constituents: Leaves contain Kaempferol glucoside, quercetain, gallic acid.

Fruit Kernel contains Proteins; minerals as Calcium, phosphorus, iron, magnessium, sodium, copper, potassium, sulpher, chlorine, thiamine, niacin, riboflavin, vitamin C.

Medicinal Properties: Root – Acrid, astringent, constipating, cooling, depurative.

Leaves – Aphrodisiac, cooling, depurative, digestive, expectorant, purgative.

Fruits – Anti-inflammatory, aphrodisiac, cooling, depurative, diuretic, emollient, expectorant, febrifuge, laxative, sour, stomachic, sweet.

Medicinal Uses: Roots – Diarrhoea, leprosy, skin diseases.

Leaf – Bronchitis, burning sensation, constipation, cough, dyspepsia, flatulence, leprosy, skin diseases.

Fruits – Asthma, burning sensation, cardiac debility, constipation, cough, fever, inflammations, leprosy, skin diseases, ulcers.

77. *Butea monosperma* (Lamk.) Taub. (Fabaceae)

Syn.: **–** *B. frondosa* Koenig ex Roxb.

Vern.: Palash.

Distribution: In Sal and mixed forests and Wastelands. Balaghat, Bhopal, Chhatarpur, Damoh, Hosahangabad, Indore, Jabalpur, Mandla, Rewa, Sagar, Seoni, Shahdol, Shivpuri, Sidhi, W. Nimar.

Botany: Medium-sized trees, with 3-foliolate large leaves and orange, showy flowers in racemes. Pods grey-silky.

Flowering & Fruiting: March – June.

Medicinal Parts: Bark, flowers, seeds, gum.

Chemical constituents: Plant contains Fats, oils, Mucilage, protein, Resin, Butin, Glucoside, Butein, Isomonospermoside, Coreopsin, Sulphruein, Palastirin, butrin, Isobutrin.

Gum contains Thiamine, Riboflavin, fatty acids, myristi acid, palmitic & stearic acid.

Medicinal Properties: Bark – Acrid, anthelmintic aphrodisiac, appetiser, bitter, hot, laxative.

Gum – Astringent.

Flower – Acrid, astringent, bitter, hot, sweet.

Fruits – Anthelmintic, aperient, digestive, dry, hot.

Seed – Anthelmintic, aperient, digestive, dry, hot.

Medicinal Uses: Root – Elephantiasis, nightblindness.

Bark – Bone fractures, dysentery, hydrocele, piles, tumour, ulcer.

Leaf – Eye diseases

Gum – Cough, dysentery, stomach troubles.

Flower – Burning sensation, eye diseases, gout, leprosy, skin diseases, strangury.

Fruits – Abdominal pain, piles, scorpion sting, skin diseases, tumours, urinary discharges.

Seed – Abdominal pain, piles, scorpion sting, skin diseases, tumours, urinary discharges.

Ayurvedic preperations: *Palashbijadi churna, Palashksar ghrit, Palashadi kwath, Calcury, Krmimudgara rasa, Palasa puspasava.*

78. *Caesalpinia bonduc* (L.) Taub. (Caesalpiniaceae)

Syn.: *C. bonducella* Fleming.; *C. crista* L.; *C. nuga* Ait. f.; *Guilandina bounducella* L.

Vern.: Kantak karanj.

Distribution: On forest edges and roadsides. Damoh, Indore, Jabalpur, Mandla, Ratlam, Rewa, Sagar, Seoni, Shivpuri, Sidhi, Tikamgarh.

Botany: Armed shrubs, pinnae 6-8 pairs, with a pair of hooked stipular spines; leaflets 6-9 pairs, elliptic-oblong, obtuse. Racemes often panicled; pods shortly stalked, oblong, densely armed on the faces with wiry prickles, 1-2 seeded.

Flowering & Fruiting: June – February.

Medicinal Parts: Root, leaves, seeds.

Chemical constituents: Seed contains Fatty oils, starch, sucrose, caesalpins, bonducellin, fatty acids, phytosterinin, saponins, phytosterols, amino acids, a bitter glucoside – bonducin.

Medicinal Properties: Leaf – Anthelmintic.

Flower – Bitter.

Fruits – Acrid, anthelmintic, aphrodisiac, astringent.

Seed – Anthelminitic, antiperiodic, styptic.

Medicinal Uses: Leaf – Elephantiasis, small pox.

Fruit – Leucorrhoea, piles, urinary discharges, wounds.

Seed – Colic, hydrocele, inflammation, leprosy, malaria, skin diseases.

Ayurvedic preperations: *Bijai churna, Karanjadi ghrit, Karanjadi churna* & *Visham javarghani vati, Lata karanja vati, Latakaranja bija taila.*

79. *Cajanus cajan* (L.) Millsp. (Fabaceae)

Syn.: *C. indicus* Spreng.

Vern.: Arahar.

Distribution: In Cultivation. Bhopal, Indore, Jabalpur, Mandla, Ratlam, Sagar, Shahdol, Shivpuri, Sidhi.

Botany: An erect shrub. Leaflets 3, oblong-lanceolate. Flowers yellow, in corymbose racemes. Pods narrowed at the ends, torulose.

Flowering & Fruiting: August-March.

Medicinal Parts: Leaves, Seeds.

Chemical Constituents: Leaves contain n-Alkanes, Simple esters, Free alcohols, Lupeol, Phytosterols as β-Sitostrerol, Stigmasterol, Campesterol, Cholesterol; Isoflavone as Cajanol; Salicylic acid, Linoleic acid, Oleic acid.

Medicinal Properties: Leaves – Anti-inflammatory, Astringent, Anodyne, Cooling, Diuretic, Laxative, Sweet.

Seeds – Acrid, Alexeteric, Anthelmintic, Astringent, Cooling, Constipating, Febrifuge, Resolvent, Sweet.

Medicinal Uses: Leaves / Seeds – Bronchitis, Cardiac diseases, Cough, Fever, Haemorrhoids, Intestinal Worms, Tumours, Vomiting.

80. *Calotropis gigantea* (L.) Ait. f. (Asclepiadaceae)

Vern.: Safed aak.

Distribution: Common in wastelands and along bundh of cultivated fields. Chhatarpur, Damoh, Hosahangabad Mandla, Seoni, Shahdol.

Botany: Shrubs or small trees, with elliptic-oblong or obovate-oblong leaves. Corolla-buds ovoid; corona-lobes hairy, shorter than staminal column.

Flowering & Fruiting: March-May.

Medicinal Parts: Root, Leaves, Flowers, Milk.

Chemical constituents: Plant contains Mudarol, akundarol, uscharidin calotropin, frugoside, calotroposides – A to G, calotroplupenyl acetate – A, leupeol acetate – B, gigantursenyl acetate – A, gigantursenyl acetate – B.

Medicinal Properties: Plant – Alexipharmic, anthelmintic, Purgative.

Root bark – Diaphoretic.

Flower – Analgesic, anthelmintic, astringent, bitter, sweet.

Milk – Bitter, purgative.

Medicinal Uses: Plant – Diseases of the spleen, the liver, the abdomen, Leprosy, leucoderma, piles, tumour, ulcer.

Root bark – Asthma, syphilis.

Flower – Ascites, inflammation, tumour.

Milk – Ascites, diseases of the abdomen, leucoderma, tumour.

81. *Calotropis procera* (Ait.) R. Br. (Asclepiadaceae)

Vern.: Aak.

Distribution: Common in wastelands, along bundh of cultivated fields and roadsides. Balaghat, Chhatarpur, Damoh, Hosahangabad, Jabalpur, Panna, Rewa, Sagar, Satna, Seoni, Shahdol, Sidhi.

Botany: Shrubs or small trees, with ovate or ovate-oblong leaves. Corolla-buds hemispherical; corona-lobes glabrous, equalling or longer than staminal column.

Flowering & Fruiting: Throughout the year.

Medicinal Parts: Root, leaves, flowers, latex.

Chemical constituents: Plant contains Cardiac steroid glycoside – terpenoids, resin, glycosides, carbohydrates, aluminium, Iron, magnesium, sodium.

Medicinal Properties: Analgesic, anthelmintic, anti – inflammatory, depurative, digestive, emetic, expectorant, laxative, resolvent, stomachic.

Medicinal Uses: Asthma, cough, dropsy, eczema, inflammation, piles, rheumatism, stranguary, ulcer.

Ayurvedic preperations: *Arkavaleha, Arka taila, Arkesvara, Arkadi churna, Ekangavira, Sutikabharana, Mahavatavidhvamsa, Pravala pancamrta, Arkakolakanda vati.*

82. *Canavalia gladiata* (Jacq.) DC. (Fabaceae)

Syn.: *C. ensiformis* Sensu Baker non DC.

Vern.: Bansem.

Distribution: On roadside bushes. Chhatarpur, Gwalior, Indore, Mandla, Satna, W. Nimar.

Botany: A perennial twiner, with white or lilia flowers and sword shaped flattened pods.

Flowering & Fruiting: August – February.

Medicinal Parts: Fruits.

Chemical Constituents: *Seeds contain* Crude Protein 29%, Crude Fiber 6%, Starch 37%, Ether extract 1.5%, Nitrogen free extract 62% and total Sugar 7.5%. The amino acid contents in the protein fraction were: Lycine 6.5%, Histidine 4%, Arginine 6%, Tryptophan 2%, Treoline 6%, Cistine 0.8%, Valine 6%, Methionine 0.5%, Isoleucine 5%, Leucine 9%, Tyrosine 2% and Phenylalanine 5%.

Medicinal Properties: Fruits – Appetiser, Astringent, Cooling, Digestive, Sweet, Vulnerary.

Medicinal Uses: Fruits – Anorexia, Burning Sensation, Dyspepsia, Ulcers, Wounds.

83. *Cannabis sativa* L. (Cannabinaceae)

Vern.: Bhang.

Distribution: Ocasionally in wastelands and along roadsides near habitations, Chhatarpur, Indore.

Botany: Dioecious, aromatic, woody herbs with 3 to 9 – foliolate leaves. Flowers white or cream, males in panicels, female's axillary, solitary forming spikes.

Flowering & Fruiting: March-November.

Medicinal Parts: Leaves, the part of male flowering twig resin.

Chemical constituents: Plant contains Cannabinoids, cannabispirans, alkaloids, cannabig lendol, cannabidiolic acid, resin (Cannabinone), gum, sugars, calcium, phosphate, and volatile oil.

Medicinal Properties: Analgesic, antispasmodic, aphrodisiac astringent, digestive, diuretic, narcotic, stiumulant, stomachic.

Medicinal Uses: Leaves – Anthelmintic, cuts, ear trouble, skin diseases, wounds.

Flowers – Diarrhoea, dysnetery, gonorrhoea.

Ayurvedic preperations: *Madnanand modak, Jati-phaladi churna, Lai churna, Majun-phalakser.*

84. *Canscora decussate* (Roxb.) Schult. (Gentianaceae)

Vern.: Sankhaphuli.

Distribution: Along roadsides in open forests. Hosahangabad.

Botany: Dichotomously branched annual, with quadrangular Stem and White flowers. *Flowering & Fruiting:* September-October.

Medicinal Parts: Whole plant.

Chemical Constituents: Loliolide, Triterpenes as Glucose Conscoradione, Friedelin, Canscoridine, Sterols as-Sitosterol, Stigmasterol, Campesterol.

Medicinal Properties: Plant – Acrid, Anthelmintic, Aphrodisiac, Appetiser, Astringent, Bitter, Emollient, Elexeteric, Depurative, Laxative, Thermogenic, Tonic, Vulnerary.

Medicinal Uses: Plant – Abdominal disorders, Ascites, Epilepsy, General debility, Inflammation, Intestinal worms, Leprosy, Leucoderma, Nervous debility Skin diseases, Tuberculosis, Ulcers.

85. *Capparis decidua* (Forsk.) Edgew. (Capparidaceae)

Vern.: Karer.

Distribution: In drier areas. Gwalior, Indore, Shivpuri.

Botany: Leafless spiny shrubs or small trees, with red or Orange-Red flowers in corymbose racemes or fascicles. Berries subglobose, scarlet when ripe.

Flowering & Fruiting: March-July.

Medicinal Parts: Roots, Fruits.

Chemical constituents: Isothiocyanate glucoside, glucocapparin, stachydrine, n-triaconate, β-carotene, β-sitotsterol, n-triacontanol, n-pentacosane, phthalic acid.

Medicinal Properties: Roots – Acrid, alexeteric, anodyne, anthelmintic, bitter, carminative, digestive, expectorant, purgative, sudorific, thermogenic, vulnerary.

Fruits – Acrid, aphrodisiac, astringent, bitter, stimulant, sweet, thermogenic.

Medicinal Uses: Root – Amenorrhoea, arthritis, asthma, boils, constipation, cough, dysmenorrhoea, dyspepsia, eruptions, flatulence, gout, hiccough, intestinal worms, lumbago, swelling, vomiting.

Fruits – Cardiac disorders.

86. *Capparis sepiaria* L. (Capparidaceae)

Distribution: *In scrub* forests. Damoh, Gwalior, Indore, Rewa, Shivpuri.

Botany: Perennial shrubs, with simple leaves & White flowers.

Flowering & Fruiting: March-December.

Medicinal parts: Root, Stem Bark, Leaf, Fruit.

Chemical constituents: Alkaloid, Glycoside, Anthocyanin, Terepene, Sterols.

Medicinal properties: *Plant* – Febrifuge, Alterative, Tonic, Anit-Inflammatory.

Medicinal uses: *Plant* – Skin diseases, Rheumatism, Fevers, Cardiac edema, Liver diseases.

87. *Capparis spinosa* L. (Capparidaceae)

Vern.: *Kabra.*

Distribution: In dry Hilly places.

Botany: Undershrubs, flowers White suffused with Red.

Flowering & Fruiting: January-September.

Medicinal parts: Root bark.

Chemical constituents: Flower buds contain Glucoside, Rutin, Rutic acid, Pectic acid, a Volatile Emetic Constituent, Saponin etc.

Medicinal properties: *Root bark* – Aperient, Tonic, Diuretic, Expectorant, Anthelmintic, Emmenagogue, Analgesic, Depurative, Anti-haemorrhoidal.

Medicinal uses: *Root bark* – Rheumatism, Paralysis, Enlarged spleen, Tubercular glands, Enlargment of Spleen & Liver.

88. *Capparis zeylanica* L. (Capparidaceae)

Syn.: *C. horrida* L. f.

Vern.: Ardanda.

Distribution: In dry deciduous forests and roadsides. Balaghat, Indore, Jabalpur, Mandla, Rewa, Satna, Sagar.

Botany: Climbing shrubs; leaves ovate or oblong, entire; stipular spines hooked; petals densely hairy inside; berries subglobose, reddish-brown, with woody pericarp; seeds subglobose.

Flowering & Fruiting: January-June.

Medicinal Parts: Root Bark, Leaves, Fruits.

Chemical Constituents: An alkaloid – α-Phytosterol, α-Mucilaginous Substance and water soluble acid. L-Stachydrine, Rutin and β-Sitosterol.

Seeds / Leaves – Thioglucoside, Glucocapparin, n-Tricontane, α-amyrine & Fixed oil.

Medicinal Properties: Root Bark **–** Appetiser, Demulcent, Sedative, Stomachic.

Medicinal Uses: Root Bark – Colic.

89. *Capsicum annuum L.* (Solanaceae)

Syn.: *C. longum* DC.; *C. grossum* Willd.; *C. condiforme* Mill.

Vern.:Lal mirch.

Distribution: Cultivated throughout the State as cash crop. Bhopal, Chhatarpur, Damoh.

Botany: Perennial herbs, with lanceolate leaves, white flowers & slender, Green red or orange fruits.

Flowering & Fruiting: Throughout the year.

Medicinal Parts: Fruits.

Chemical constituents: Fruits contain A volatile alkaloid-capsicin; a crystalline substance-capsacin; a volatile oil-solanine; fatty acid, resin.

Medicinal Properties:

Fruits – Acrid, antiperiodic, bitter, carminative, digestive, expectorant, laxative, rubefacient, stimulant, sudorific, thermogenic.

Medicinal Uses: Fruits – Arthritis, cholera, cough, dropsy, dyspepsia, fever, flatulence, gout, hoarseness, sciatica.

90. *Capsicum frutescens* L. (Solanaceae)

Syn.: *C. minimum* Roxb.

Distribution: Throughout the state.

Botany: Herbs; fruiting pedicels solitary erect; fruits ovoid-oblong, obtuse or rounded, Red 1-3 cm. long.

Flowering & Fruiting: Throughout the year.

Medicinal parts: Fruit

Chemical constituents: *Fruit contain* – Capsaicin & Dihydrocapsaicin.

Medicinal properties: *Fruit* – Acrid, Stimulant, Carminative, Rubifacient.

Medicinal uses: *Fruit* – Dyspepsia, Stomachic, Indigestion, Diarrhoea.

91. *Cardiospermum halicacabum* L. (Sapindaceae)

Vern.: Kanphuta.

Distribution: Climbing on bushes and hedges along roadsides, forest-clearings, grasslands and in dry deciduous forests. Also common in cultivated fields. Balaghat, Betul, Chhatarpur, Damoh, E. Nimar, Guna, Gwalior, Hosahangabad, Indore, Jabalpur, Jhabua, Mandla, Sagar, Satna, Seoni, Tikamgarh.

Botany: Herbaceous climbers, with 5-fid axillary tendrils and bi-or trinate leaves. Flowers white, in cymes on terminal tendrils.

Flowering & Fruiting: Throughout the year.

Medicinal Parts: Roots, Leaves, Seeds.

Chemical constituents: Seed – Cyanogenic glycoside, essential oil.

Medicinal Properties: Root – Diaphoretic, diuretic, emetic, laxative.

Leaves – Rubefacient.

Seeds – Diaphoretic.

Medicinal Uses: Root – Menorrhoea, arthritis, fever, lumbago, strangury.

Leaves – Arthritis.

Seeds – Arthritis, fever.

92. *Careya arborea* Roxb. (Borringtoniaceae)

Distribution: Growing on hill slopes and along the sides of streams. Balaghat, Mandla, Seoni, Sidhi.

Botany: A small to medium sized deciduous tree with yellowish-white flowers and berries fruit.

Flowering & Fruiting: March-July.

Medicinal Parts: Bark, leaves, flowers, fruits.

Medicinal Properties:

Bark – Acrid, alexeteric, anthelmintic, astringent, bitter, expectorant, thermogenic.

Fruit – Acrid, astringent, aromatic, anaphrodisiac.

Medicinal Uses: Bark – Bronchitis, catarrh, colic, cough, dysentery, dyspepsia, leucoderma, tumours.

Fruit – Dyspepsia.

93. *Carica papaya* L. (Caricaceae)

Vern.: Papita.

Distribution: Commonly cultivated, almost throughout State.

Botany: A rapidly growing tree, up to 7 m. tall, with milky sap. Leaves palmately-lobed and flowers creamy yellow; female flowers in short clusters and males in long drooping panicles.

Flowering & Fruiting: Almost throughout the year.

Medicinal Parts: Fruits, Latex.

Chemical constituents: Fruits pulp contains Sucrose, invert sugar, papain, malic acid, salts of tartaric acid & citric acid, pectins, carotenoid, vitamins.

Latex contains Enzymes papain & chymopapain.

Medicinal Properties: Fruits – Acrid, anodyne, anthelmintic, anti-inflammatory, antifungal, aphrodisiac, appetizer, bitter, carminative, demulcent, digestive, stomachic, thermogenic.

Latex – Anodyne, anthelmintic, febrifuge, sudorific.

Medicinal Uses: Fruits – Bronchitis, cough, dyspepsia, inflammations, intestinal worms, psoriasis, ringworm, skin diseases.

Latex – Amenorrhoea, constipation, dysmenorrhoea, dyspepsia, leprosy, skin diseases.

94. *Carissa carandas* L. (Apocynaceae)

Syn.: *C. cognesta* Wt.

Vern.: Karonda.

Distribution: Common along the edges of forests also under cultivation. Damoh, Jabalpur, Sagar.

Botany: Erect or scandent shrubs, with paired spines at each node. Leaves elliptic-oblong to ovate. Flowers white, in corymbose cymes. Berries elliptic-oblong, fleshy.

Flowering & Fruiting: May-October.

Medicinal Parts: Root, Fruits.

Chemical Constituents: Root contain – β-sitosterol, a Sesquiterpene as Carissone: a Lignancarinol, Odoroside-H, Digitoxigenin, Glucose, D-digitalose, Lupeol, Ursolic acid, Methyl ursolate.

Fruits contain – Triterpene alcohol as Carissol, Glucose, Galactose, Serine, Glutamine, Alanine, Valine, Glycine, Carissic acid, Phenylalanine.

Medicinal Properties: Leaves – Febrifuge.

Roots – Anthlemintic, Antiscorbutic, Stomachic.

Ripe fruits – Antiscorbutic, Appetiser, Cooling, Sweet.

Unripe fruits – Antipyretic, Aphrodisiac, Appetiser, Astringent, Bitter, Constipating, Sour, Thermogenic.

Medicinal Uses: Roots – Intestinal worms, Pruritus, Scabies, Stomach disorders.

Ripe fruits – Burning sensation, Skin diseases, Scabies, Pruritus.

Unripe fruits – Anorexia, Diarrhoea, Intermittent fever.

95. *Caryota urens* L. (Arecaceae)

Vern.: Mari.

Distribution: Planted. Bhopal, Indore.

Botany: A tall, feather-leaved palm, with smooth shiny trunk and pendulous spadix inflorescence.

Flowering & Fruiting: October-March.

Medicinal Parts: Tender bud, leaves & nuts.

Chemical constituents: Leaves contain Epicatechin, Gallo-catechin and epicatechingallate.

Medicinal Properties: Tender leaves – Cooling, sweet.

Medicinal Uses: *Fruits* – Fatigue, hyperdisia.

96. *Cassia fistula* L. (Caesalpiniaceae)

Syn.: C. rhombifolia Roxb.; C. excelsa H. B. & K.

Vern.: Amaltas.

Distribution: In Sal and mixed forests and roadsides. Balaghat, Bhopal, Chhindwada, E. Nimar, Hosahangabad, Indore, Jabalpur, Khandwa, Mandla, Panna, Raisen, Sagar, Satna, Sehore, Seoni, Shahdol, Shivpuri, Sidhi.

Botany: Deciduous tree; branchlets glabrous; leaves eglandular; leaflets 4-8 pairs,ovate, silvery, pubescent beneath when young; flowers in lax racemes; pods long, cylindrical, nearly straight, smooth, shining, brown-black.

Flowering & Fruiting: *March – May.*

Medicinal Parts: Whole plant.

Chemical constituents: Fruit contains Glycosides, carbohydrates, resin, tannins, magnessium, calcium, potassium, iron & other compounds are fistucacidin, beta-sitosterol, n-tricontanol, leucopelavgonidin, flavonoids, glucosides, gum, and volatile oil.

Medicinal Properties: Root – Astringent, cooling, febrifuge, purgative, tonic.

Bark – Anthelmintic, diuretic, emetic, febrifuge, laxative.

Leaves – Antiperiodic, depurative, laxative.

Flower – Acrid, bitter, cooling, demulcent, emollient, expectorant.

Fruit – Abortifacient, anti-inflammatory, antipyretic, cooling, diuretic, emollient, ophthalmic, purgative, sweet.

Medicinal Uses: Root – Burning sensation, leprosy, skin disease, syphilis, tuberculous glands.

Bark – Boils, colic, constipation, diabetes, dyspepsia, fever, leprosy, pustules, ringworm.

Leaves – Fever, leprosy, skin disease, ulcers.

Flowers – Burning sensation, bronchitis, cough, pruritus, skin disease.

Fruit – Burning sensation, cardic disorders, colic, fever, gout, inflamation, jaundice, pruritus, rheumatism, Skin disease.

Ayurvedic preperations: *Aragvadhadi-kshar, Aragvadhadi-taila, Aragvadhadi-leh, Aragvadharista, Aragvadhadi-Sutravati, Aragvadha kwatham.*

97. *Cassia sophera* L. (Caesalpiniaceae)

Syn.: *Senna sophera* (L.) Roxb.

Vern.: Kasunda.

Distribution: On edges of mixed forests. Bhopal, Chhattarpur, Mandla, Sagar.

Botany: Shrubs; leaflets 5-10 pairs; flowers yellow; pods turgid.

Flowering & Fruiting: November-February.

Medicinal parts: Bark, Leaves, Seeds, Root & Root Bark.

Chemical constituents: Emodin and Chrysophani acid.

Seed contain tannic acid, Mucilage, Fatty oil Emodin & a Toxalbumin, Rhamnetin-3-0-B-D Glucoside along with Chrysophanol from flowers.

Medicinal properties: *Leaves, Bark, Seed* – Cathartic.

Medicinal uses: *Leaves* – used externally in Ringworm.

Decoction of plant: Used in acute bronchitis.

98. *Catharanthus roseus Don* (Apocynaceae)

Syn.: *Vinca rosea* L.; *Lochnera rosea* Reichb.

Vern.: Sadabahar.

Distribution: Common in sand wastelands and along roadsides. Balaghat, Betul, Chhatarpur, Hosahangabad, Indore, Satna, Seoni, Sidhi.

Botany: An erect herb, with simple opposite leaves; white or rosy flowers.

Flowering & Fruiting: Round the year.

Medicinal Parts: Whole plant.

Chemical constituents: Plant contains Indole alkaloids – vincamine; BisIndole alkaloids–vinblastine. Vincristine; Aracidic acid, Behenic acid, Lauric acid, Linolenic acid, oleic acid, palmitic acid, Stearic acid (Seeds).

Stem contains β – Sitosterol, Ursolic acid.

Medicinal Properties: Acrid, astringent, bitter, laxative, stomachic, toxic.

Medicinal Uses: Diabetes, haemorrhagic diseases, leucomea, menorrhagea.

99. Cayratia carnosa (Wall.ex Wight) Gagnep. (Vitaceae)

Syn.: *Vitis carnosa* L.; *Vitis trifolia* L.

Vern.: Amal-bel.

Distribution: Balaghat, Damoh, Dhar, Gwalior, Hosahangabad, Indore, Mandla, Panna, Rewa, Sagar, Seoni, Shivpuri, Sidhi.

Botany: Herbaceous climbers, with 3 – foliolate leaves and greenish – white flowers in axillary or pseudo, terminal cymes. Berries obovate – globose, dark purple to black when ripe.

Flowering & Fruiting: July-October.

Medicinal Parts: Whole plant.

Medicinal Properties: Astringent, diuretic, sour.

Medicinal Uses: Cardiac disorders, dropsy, fever, haemorrhoids, neuralgia, splenopathy, strangury, tumors, ulcer, wounds.

100. *Celastrus paniculatus Willd.* (Celastraceae)

Syn.: *C. montana* Roth.; *C. multiflora* Roxb.

Vern.: Malkangani.

Distribution: Balaghat, Damoh, Hosahangabad, Panna, Satna, Seoni, Shivpuri, Sidhi.

Botany: Unarmed, shrubby climbers, with ovate-elliptic leaves and greenish yellow flowers in terminal panicles. Ovary free from the disc.

Flowering & Fruiting: November– February.

Medicinal Parts: Bark, leaves, seeds.

Chemical constituents: Seed contains Concentrate reddish-brownish, bitter and pungent oil, Resin, Tannin, celastrin and paniculatin.

Medicinal Properties: Bark – Abortifacient, depurative.

Leaf – Antidote.

Seed – Acrid, anti-Inflammatory, aphrodisiac, appetiser, bitter, digestive, diuretic, emetic, emollicent, expectorant, febrifuge, intellect, laxative, promoting, stimulant, thermogenic.

Medicinal Uses: Bark – Brain tonic.

Seed – Abdominal disorders, amenorrhoea, asthma, cardiac debility, dysenorrhoea, fever, inflammation, leucoderma, leprosy, paralysis, pruritus, skin diseases.

Ayurvedic preperations: *Jyotishmati taila, Mall taila, Marichyadi taila, Karanjadi yog, Laghu-vis-garva taila, Rogan malkangni, Jyotismati churna.*

101. *Celosia argentea* L. (Amaranthaceae)

Syn.: *C. cristata* L.

Vern.: Safed murga.

Distribution: Common as weed in cultivated fields, wastelands, along roadsides and in dried up river beds. Balaghat, Chhatarpur, Damoh, E. Nimar, Gwalior, Hosahangabad, Indore, Mandla, Panna, Sagar, Satna, Seoni, Shivpuri, Sidhi, Tikamgarh, W. Nimar.

Botany: Erect annuals. Leaves linear-lanceolate or ovate-elliptic. Flowers white or pink. Seeds lenticular, black, shining.

Flowering & Fruiting: August-December.

Medicinal Parts: Leaves, seeds.

Chemical constituents: Seed contain Saponin of hyaluronic acid.

Medicinal Properties: Aphrodisiac, astringent, cooling, nutritive, spasmolytic.

Medicinal Uses: Seeds – Diarrhoea, dysuria, seminal diseases.

102. *Centella asiatica* (L.) Urban (Apiaceae)

Syn.: *Hydrocotyle asiatica* L.

Vern.: Mandukaparni.

Distribution: Balaghat, Hosahangabad, Indore, Sidhi.

Botany: Creeping herbs, with broadly suborbicular leaves and pink or red flowers in fasciculate umbels. Fruits ovoid.

Flowering & Fruiting: May – November.

Medicinal Parts: Whole plant.

Chemical constituents: Plant contains Alkaloids-Brahmi, Herpestin, Bacosides A & B In Bacoside-A: Arabnosil, glucose, Arabnose, Becogenine, Betulic acid, D-Manitol, Stigmasterol, β-Sitosterol & Tannins.

Leaves contain Glucoside, Volatile oil.

Dry parts contain Centoic acid, centalic acid.

Whole plant also contains Pectic acid, valerin, resin, ascorbic acid.

Medicinal Properties: Acrid, antileprotic, bitter, cardiotonic, carminative, cooling, diuretic, febrifuge, nervine tonic, stomachic, sweet.

Medicinal Uses: Abdominal disorders, amentia, asthma, bronchitis, cardiac debility, epilepsy, fever, hiccough, hoarseness, insomnia, leprosy, strangury.

Ayurvedic preperations: *Brahmi rasayana, Brahmi taila, Brahmi sharbat, Brahmi ghirit, Brahmi panak, Sarasvatarishta, Sarasvat ghirit, Brahmi halwa, Brahmi malhar, Majum brahmi, Astanga ghirit.*

103. *Centipeda minima* (L.) A. Br. & Aschers (Asteraceae)

Syn.: *C. orbicularis* Lour.

Distribution: Common in moist places. Damoh, Mandla, Seoni, Sidhi.

Medicinal parts: Leaves.

Chemical constituents: *Plant* – Lupeol, Hexacosanol, β-Sitosterol, Stigmasterol, Centipedaoic acid, Flavons.

Medicinal properties: *Seed* – Vermifuge.

Leaf – Hot, Sharp, Dry, Acrid, Anthelmintic.

Medicinal uses: *Leaf* – Leucoderma, Diseases of the Blood and the Skin, Hysteria.

Seed – Headache, Colds in the head, Paralysis, Pains in Joints.

104. *Chenopodium album* L. (Chenopodiaceae)

Vern.: Bathua.

Distribution: Common in cultivated fields, along river banks and in wastelands. Chhatarpur, Damoh, Hosahangabad, Jabalpur, Mandla, Panna, Sagar, Satna, Sidhi.

Botany: Herbs, with oblong-rhomboid, somewhat fleshy leaves. Flowers pale-green, born in clusters in paniculate, ebracteate spikes; perianth-segements without any swelling on the back. Seeds rugose.

Flowering & Fruiting: November– February.

Medicinal Parts: Whole plant.

Chemical constituents: Plant contains A volatile oil, which contains Ascaridol; saponin, Vitamin – C, Magnessium, Phosphate, Carotene.

Medicinal Properties: Acrid, anthelmintic, aphrodisiac, carminative, digestive, diuretic, laxative, Sweet, tonic.

Medicinal Uses: Cardiac disorders, dyspepsia, flatulence, general debility, helminthiasis, Peptic ulcer, seminal weakness, strangury.

105. *Chlorophytum tuberosum* Bak. (Liliaceae)

Vern.: Safed musli.

Distribution: In Sal forests and in open places among grasses. Hosahangabad.

Botany: Annual herbs, with many cylindric root – fibres ending in ellipsoid tubers. Leaves linear – lanceolate. Flowers white, in dense racemes. Capsules obovoid transversely veined.

Flowering & Fruiting: July – August.

Medicinal Parts: Root, tubers.

Chemical constituents: Root contains Steroids, resin, phenolics, tannins, carbohydrates, calcium, magnessium and potassium.

Medicinal Properties: Root and tuber – Aphrodisiac, cooling, diuretic, tonic.

Medicinal Uses: Root and tuber – Diarrrhoea, dysentery, general debility, gonorrhoea, leucorrhoea, seminal debility.

106. *Chrozophora rottleri* (Geis.) A. Juss. ex Spreng. (Euphorbiaceae)

Syn.: C. plicata Hook. F.

Distribution: Common in wastelands, dried ditches, roadsides, river beds & along the railway tract. Damoh, E. Nimar, Mandla.

Botany: Stellately hairy herbs, branched in the upper region. Leaves entire or 3-lobed, ovate to suborbicular. Sepals of female flowers triangular; stigma red. capsules purple.

Flowering & Fruiting: March-May.

Medicinal Parts: Seeds.

Medicinal Properties: The plant has Corrosive Properties, Drastic, Emetic.

Medicinal Uses: Seeds – Cathartic.

107. *Cicca distichia* L. (Euphorbiaceae)

Syn.: *C.acida* (L.)Merr.; *Phyllanthus distichus* Muell-Arg.; *P. acidus* Skeels.; *Averrhoea acida* L.

Vern.: Harfarebari.

Distribution: Commonly cultivated in gardens & parks. Bhopal, Indore.

Botany: Tree, with erect branches at top; branchlets slender. Leaves distichous thin. Obliquely ovate acute, base rounded, tip acute. Flowers very small in slender racemes. Fruits yellow-green, globose, fleshy, depressed both ends.

Flowering & Fruiting: October-April.

Medicinal Parts: Root, Fruits, Seeds.

Chemical Constituents: Root bark contains Tannin, Saponin, Gallic acid and a Crystalline substance, β-amyrin, Phyllanthol.

Medicinal Properties: Fruits – Appetiser, Astringent, Sour, Tonic.

Medicinal Uses: Fruits – Biliousness, Blood purifier Bronchitis, Constipation, Piles, Vomiting.

108. *Cicer arietinum* L. (Fabaceae)

Distribution: Commonly cultivated as a cold weather crop for its seeds used as a pulse. Bhopal, Chhindwara, Hosahangabad, Indore, Mandla, Sagar, Shivpuri, Sidhi.

Botany: Much branched herbs, with pinnate leaves and turgid-oblong fruits.

Flowering & Fruiting: November – March.

Medicinal Parts: Leaves, Seeds.

Chemical constituents: Leaves contain Vanillic, P-hydroxybenzoic acid; oxalic acid, malic acid, arginine, tryosine, lycine, cystine, tryptophane; carotenoids & vitamins.

Medicinal Properties: Leaves – Abortifacient, purgative.

Seed – Aphrodisiac, cooling, expectorant.

Medicinal Uses: Leaves – Bronchitis, Dyspepsia.

Seed – Bronchitis, burning sensation, inflammations, leprosy, skin diseases.

109. *Cinnamomum tamala* (Buch.-Ham.) Nees. (Lauraceae)

Vern.: Tejpat.

Distribution: Planted in gardens. Hosahangabad.

Botany: A medium sized, evergreen, scented, about 25 feet high, flowers yellowish in panicles.

Flowering & Fruiting: July-October.

Medicinal Parts: Leaves.

Chemical constituents: Leaves contain – Essential oil, eugenol and isoeugenol.

Medicinal Properties: Bark – Acrid, alexeteric, aphrodisiac, aromatic, astringent, bitter, carminative, diuretic, expectorant, febrifuge, stimulant, sweet.

Medicinal Uses: Bark – Asthma, bronchitis, diarrhoea, fever, flatulence, nausea, vomiting.

110. Cissampelos pareira L. (Menispermaceae)

Syn.: *C. hexandra* Roxb.; *C. convolvulacea* Willd.; *C.hernadifolia* Wall.

Vern.: Akauadi.

Description: Along road sides and in the forests. Common throughout the State.

Botany: Twining, dioecious shrubs, with orbicular or reniform leaves, greenish yellow flowers and subglobose fruits which become red on ripening.

Flowering & Fruiting: June-January.

Medicinal Parts: Root, bark.

Chemical constituents: Root & its bark contains Alkaloids, cissampareine, berberine, cissampeline, pelosine, hayantin, hayatinin, saponin.

Medicinal Properties: Plant – Bitter, hot.

Root – Bitter, stomachic.

Medicinal Uses: Plant – Asthma, burning sensation, dysentery, eruptions, fever, heart trouble, itching, piles, skin diseases, ulcers, vomiting.

Root – Conjunctivitis, cough, diarrhoea, dropsy, dyspepsia, scorpion sting, snakebite, stomach pain.

Ayurvedic preperations: *Ambhashtadi churna, Gangadhar churna, Kutajastak kwath, Kutaj-ghanvati, Majun jograj guggul.*

111. *Citrus aurantium* L. (Rutaceae)

Vern.: Santara.

Distribution: Mandla, Shahdol.

Botany: Small trees, with White fragrant flowers and Hesperidium fruits.

Flowering & Fruiting: July-March.

Medicinal parts: Leaves, Bark, Fruits, Peel.

Chemical constituents: *Peel* – Flavonoid; Auranetin.

Essential oil – d-Limonene, Monoterpenes, Sesquiterpenes, Flavonoids, Triterpenoids, Alkaloids: Synerphrine, Citral, Citronellal, β & α-Pinene, Limonene, α-Phellandrene, Decanal, α-Terpineol, Geraniol, Palmitic acid.

Fruits – Hesperidin, Neohesperidin, Limonene, Naringin, poncirin & umbelliferone.

Medicinal properties: *Fruits* – Laxative, Stomachahic.

Peel oil – Antifungal.

Medicinal uses: *Leaves* – Headache, Abdominal pain, Urinary tract infections.

Peel oil – Bronchitis.

Bark – Sunstroke, Urinary tract infections.

112. *Citrus limon* (L.) Burm. f. (Rutaceae)

Distribution: Bhopal, Damoh, Sagar.

Flowering & Fruiting: May-June.

Botany: Thorny trees, with ovate-elliptic leaves, white flowers tinged with pink-purple and oval or oblong fruits.

Flowering & Fruiting: April-May.

Medicinal Parts: Fruits.

Taste: Sour.

Chemical constituents: Oil from peel contains d-limonene, camphene, linalool.

Peel contains Bitter principle, essential oil, hesperidin, citronellal, bergamotine, eriocitrin, apigenin, luteolin, chrysoseriol, quercetin, isorhamnetin, limocitin, limocitriol.

Seeds contain Bitter principles, obacunone, limonin.

Roots contain Xanthyletin, stigmasterol, β-sitosterol.

Medicinal Properties: Fruits – Anthelmintic, antiseptic. carminative, digestive, laxative, stimulant, sour, stomachic, thermogenic.

Medicinal Uses: Fruits – Bronchitis, colic, constipation, cough, dyspepsia, fatigue, flatulence, helminthiasis, scabies, vomiting.

113. *Citrus maxima* (Burm.) Merr. (Rutaceae)

Syn.: *C. decumana* L.; *C. grandis* (L.) Osbeck.; *C. paradisi* Macf.

Vern.: Chakotra.

Distribution: Sagar.

Botany: Small trees; flowers creamish.

Flowering & Fruiting: January-December.

Medicinal parts: Leaves & Fruits.

Chemical constituents: Naringin.

Oil from Peel – d-Limonene, α-Pinene, Linalool.

Medicinal properties: *Fruit* – Nutritive, Cardio tonic, Refrigerant.

Medicinal uses: Leaves – Convulsive cough, Epilepsy.

114. *Citrus medica* L. (Rutaceae)

Vern.: Bara nimbu.

Distribution: Hoshangabad.

Botany: Large shrubs or small trees, with axillary spines, white flowers and globose fruits.

Flowering & Fruiting: January-December.

Medicinal Parts: Whole plant.

Chemical constituents: α-pinene, β-pinene, myrcene, camphene, limonene, α-phellandrene, 3-carene, P-cymene, γ-terpinene, terpinolence, nonylaldehyde, citronellal oil.

Medicinal Properties: Roots – Anthlemintic, diuretic, laxative.

Buds & Flowers – Astringent, antiemetic, appestiser. stimulant.

Fruits – Anthelmintic, astringent, carminative, digestive, emollient, refrigerant, sour, stimulant, sweet.

Seeds – Anti-inflammatory, stimulant.

Medicinal Uses: Root – Colic, constipation, flatulence, helminthiasis, strangury, tumours, vomiting.

Buds & Flowers – Abdominal disorders, dysmenorrhoea, flatulence, leprosy, tumours, vomiting.

Seeds – Amenorrhoea, dysmenorrhoea, inflammations, skin diseases.

115. *Clematis triloba* Heyne ex Roth. (Ranunculaceae)

Vern.: Murhari.

Distribution: As undergrowth in mixed deciduous forests. Hosahangabad, Narsimhpur.

Botany: An extensive climber, with opposite leaves and white flowers in axillary corymbose panicles.

Flowering & Fruiting: **October-December.**

Medicinal Parts: Roots.

Chemical Constituents: Anemonin, Alcoholic extract of plant is hypotensive and spamolytic.

Medicinal Properties: Leaves – Acrid, Alterative, Astringent, Bitter, Hot, Laxative, Sedative, Stimulant, Sweet.

Medicinal Uses: Leaves – Blood diseases, Syphilis, Scrofula, Leprosy, Chronic fevers, Heart troubles, Boils, Itching.

116. *Cleome gynandra* L. (Cleomaceae)

Syn.: *Gynandropsis gynandra* DC.

Vern.: Hulhul.

Distribution: In Waste lands. Damoh, Gwalior, Jabalpur, Mandla, Sagar, Satna, Sidhi.

Botany: Glandular hairy herbs, with 3 to 5-foliolate leaves. Flowers white, in terminal racemes. Capsules linear-cylindric.

Flowering & Fruiting: June – November.

Medicinal Parts: Root, seeds.

Chemical constituents: Plant contains Essential oil.

Seed contains Cleomin.

Medicinal Properties: Root – Hot sharp taste, stomachic.

Medicinal Uses: Root – Ascites, earache, fever, spleen enlargement, tumours, ulcers.

117. Cleome viscosa L. (Cleomaceae)

Syn.: *C. icosandra* L.

Vern.: Hulhur.

Distribution: In Waste lands. Balaghat, Dhar, Hosahangabad, Indore, Jabalpur, Mandla, Raisen, Rewa, Satna, Sidhi, Tikamgarh.

Botany: Glandular hariy herbs, with 3 to 5-foliolate leaves. Flowers yellow, in leafy racemes; stamens more than 10. Capsules cylindric, beaked.

Flowering & Fruiting: June – September.

Medicinal Parts: Whole plant.

Chemical constituents: Seed contain Viscosic acid & visosin.

Medicinal Properties: Plant – Acrid, anthelmintic, sudorific, thermogenic.

Root – Antiscorbutic, stimulant, vermifuge.

Leaves – Digestive.

Seed – Anthelmintic, cardiac stimulant, carminative, constipating, febrifuge.

Medicinal Uses: Root – Otaglia.

Seed – Cardiac disorders, diarrhoea, dyspepsia, fever.

118. *Clerodendrum inerme* (L.) Gaertn. (Verbenaceae)

Vern.: Choti Arni.

Distribution: Bhopal, Chhatarpur, Indore.

Botany: A glabrous shining shrub. Leaves subsessile, variable in size. Flowers white with purple-red filaments in umbelled axillary cymes. Drupes pyriform, subtended by calyx.

Flowering & Fruiting: July – April.

Medicinal Parts: Whole plant, Root & Leaves.

Chemical constituents: Leaves contain Neolignan.

Flower contains Galactose, glucose, fructose, Lysine, arginine, serine, protein, threonine & L– glutamic acid.

Stem & leaves contain Betulin, Neoclerodane, Diterpenoid, cleroinermin, β – sitosterol, clerodermic acid, acacetin, Apigenin.

Medicinal Properties: Whole plant – Bitter, blood purifier, fragrant, nutritive.

Medicinal Uses: Whole plant – Remitten fever.

Leaf – Scrofula, veneral diseases.

Root – Rheumatism.

119. *Clerodendrum phlomidis* L. f. (Verbenaceae)

Vern.: Arni.

Distribution: In Wastelands. Chhatarpur, Gwalior, Indore, Panna, Rewa, Sagar, Satna, Shivpuri, W. Nimar.

Botany: Erect or scandent shrubs, with solid stem and opposite, rhomboid – ovate leaves. Flowers white, in panicles. Fruits obovoid.

Flowering & Fruiting: October – February.

Medicinal Parts: Root, Leaves.

Chemical constituents: Plant contains Pectolinargenin, Scuttelarein, Apigenin, Clerosterol, Hispudilin, Luteolin, Clerodendrin, Cerolic acid, Cerylalcohol, Raffinose, cholesterol, β – Sitosterol.

Medicinal Properties: Root – Alexipharmic, bitter, heating, laxative, pungent, stomachic.

Medicinal Uses: Root – Anemia, constipation, diabetes, dyspepsia, fever, inflammation, piles, swelling.

120. *Clerodendrum serratum* (L.) Moon. (Verbenaceae)

Syn.: *C. dentatum* Lam.

Vern.: Bharangi.

Distribution: In Sal forests. Balaghat, Damoh, Hosahangabad, Indore, Mandla, Seoni.

Botany: A shrub, usually herbaceous but occasionally woody, with 4-angular stems. Leaves sessile or subsessile, opposite or ternate. Flowers blue, purple or white, 2-lipped, lips unequal. Drupe broadly obovoid, succulent, dark-purple when ripe.

Flowering & Fruiting: August-October.

Medicinal Parts: Roots, leaves.

Chemical constituents: Roots & leaves contain Glycosides, olianolic acid, querteroic acid, serotonic acid, manitol, saponins, orange coloured resinous substance, starch.

Medicinal Properties: Root – Acrid, anthelmintic, anti-inflammatory, antispasmodic, bitter, carminative, depurative, digestive, expectorant, febrifuge, stimulant, stomachic, sudorific.

Medicinal Uses: Root – Asthma, bronchitis, chronic, colic, consumption, cough, dropsy, dyspepsia, flatulence, fever, helminthiasis, hiccough, inflammation, leprosy, leucoderma, nasitis, skin diseases, tumours, tubercular glands.

Ayurvedic preperations: *Bharangi sura, Bharangi guda, Bharangi avaleha, Bharangyadi kvatha.*

121. *Clerodendrum viscosum* Vent. (Verbenaceae)

Syn.: *C. infortunatum* Auct.

Vern.: Bhant.

Distribution: Common along roadsides. Damoh, Hosahangabad, Rewa.

Botany: Large tomentose shrubs or small tress; branchlets 4-angled; flowers Pinkish White, in lax terminal panicles; drupe black in enlarged Pink calyx.

Flowering & Fruiting: March-June.

Medicinal parts: Leaves, Roots, Flowers.

Chemical constituents: Roots – β-Sitosterol.

Leaves – Clerosterol, Campestrol, Cholesterol, Sitosterol, Stigmasterol.

Flower – Fumaric acid, β-Sitosterol.

Medicinal properties: *Root* – Antiumour, Febrifuge, Vermifuge.

Leaves: Acrid, Anti-inflammatory, antipyretic, Antitumour, Bitter, Demulcent, Depurative, Expectorant, Febrifuge, Laxative, Thermogenic, Tonic, Vermifuge.

Medicinal uses: *Roots* – Skin diseases, Tumour.

Leaves – Bronchitis, Cough, General debility, Helminthiasis, Inflammation, Intermittent fever, Leprosy, Malaria fever, Skin diseases, Tumours.

122. *Clitoria ternatea* L. (Fabaceae)

Vern.: Visnukanta.

Distribution: On roadside bushes. Chhindwara, Hosahangabad, Indore, Jabalpur, Mandla, Raisen, Ratlam, Rewa, Sagar, Satna, Seoni, Shahdol, Shivpuri, Sidhi, W. Nimar.

Botany: Twinning herbs, with imparipinnate, 5 to 7-foliate leaves and blue or white flowers. Pods spreading.

Flowering & Fruiting: July-March..

Medicinal Parts: Root, Leaves, seeds.

Chemical constituents: Root bark contains Starch, Tannin, Resin.

Seeds contain Oil, bitter blue coloured resin.

Medicinal Properties: Root – Alexeteric, anthelmintic, aphrodisiac, bitter, depurative, diuretic, intellect promoting, laxative, ophthalmic, refrigerant, tonic.

Seed – Cathartic.

Medicinal Uses: Root – Amentia, asthma, bronchitis, burning sensation, elephantiasis, fevers, helminthiasis, inflammation, leprosy, leucoderma, Ophthalmopathy, strangury, tubercular glands, ulcers.

Leaves – Eruptions, hepatopathy, Otalgia..

Seed – Visceralgia.

Ayurvedic preperations: *Aparajita bija churna, Mula churna.*

123. *Coccinia cordifolia* Cogn. (Cucurbitaceae)

Syn.: *C. grandis* (L.) Voight.; *C. indica* W. & A.; *Cephalandra indica* Naud.

Vern.: Kanduri.

Distribution: Bhopal, Chhindwara, Jabalpur.

Botany: Slender, dioecious climbers; nodes on the mature stem annulate; leaves entire or digitately 3-5 lobed, cordate; flowers white, usually solitary axillary; fruits oblong-obtuse, green.

Flowering & Fruiting: July-May.

Medicinal Parts: Root, leaves, fruits.

Chemical constituents: Root contain β – Amyrin, Lupeol, β – sitosterol.

Fruits contain Beta-amyrin, lupeol, cucerbitacin, β – sitosterol, taraxerol, lycopene, β – carotene, crypto xanthin.

Seeds contain A fat rich in palmitic, Oleic, Linoleic acids.

Medicinal Properties: Root – Aphrodisiac, cooling.

Leaf – Astringent, bitter, cooling, sweet.

Fruit – Antipyretic, astringent cooling, depurative, expectorant, sweet.

Medicinal Uses: Root – Burning sensation, vomiting.

Leaf – Dyspepsia, inflammation, jaundice, leprosy, ulcers.

Fruits – Asthma, burning sensation, bronchitis, cough, diabetes, helminthiasis, jaundice, leprosy, skin diseases.

124. *Cocculus hirsutus* (L.) Diels. (Menispermaceae)

Syn.: *Menispermum hirsutum* L.; *C. villosus* DC.

Vern.: Jamtikibel.

Distribution: Over bushes & Small trees. Common throughout the State.

Botany: Scandent, dioecious shrubs, with cordate – lanceolate or subdeltoid leaves. Male flowers minute, yellowish – green, in panicles. Fruits reddish – purple, smooth; Seeds transversely rugose.

Flowering & Fruiting: November – May.

Medicinal Parts: Root, Leaves.

Chemical constituents: Leaves contain Trilobine & Isotrilobine.

Root contains D – Trilobine, DL – Coclaurine.

Stem contains β – Sitosterol, ginnol & mono methyl ether of Inositol.

Medicinal Properties: Root – Acrid, alexeteric, alterant, anodyne, antipyretic, aphrodisiac, bitter, carminative, demulcent, depurative, digestive, diuretic, emollient, expectorant, laxative, tonic.

Leaves – Anodyne, cooling, demulcent, expectorant, mucilaginous.

Medicinal Uses: Root – Bronchitis, buring sensation, cephalalgia, colic, cough, dyspepsia, flatulence, fractures, general debility, gout, hypertension, intermittent fevers, leprosy, poisonous bites, pruritus, skin diseases, spermatorrhoea, strangury, tubercular glands.

Leaves – Cough, eczema, gonorrhoea, neuralgia, ophthalmia.

Ayurvedic preperations: *Patalagarudi svarasa.*

125. *Cocus nucifera* L. (Arecaceae)

Vern.: Nariyal.

Distribution: Planted. Bhopal, Indore.

Botany: A tall palm tree, with subglobose drupe. Drupe Green or Yellowish; pericarp fibrous; endocarp bony; albumen lining the endocarp, the large cavity filled with a sweet fluid.

Flowering & Fruiting: Throughout the year.

Medicinal Parts: Roots, Inflorescence, Seeds (Shell, kernel, water, oil).

Chemical constituents: Kernel contain Mannan, galactomannan, enzymes, amino-acids, proteins.

Medicinal Properties: Root – Anthlemintic, astringent, diuretic.

Juice of the young spadix – Aperient, aphrodisiac, diuretic, intoxicating, refrigerant, sweet.

Shell – Cooling, diuretic.

Kernel – Aphrodisiac, appetizer, cooling, laxative, sweet.

Water – Anthlemintic, aperient, aphrodisiac, cooling, digestive, diuretic, sweet.

Oil – Aphrodisiac, appetiser, digestive, sweet.

Medicinal Uses: Root – Bronchitis, helminthiasis, strangury.

Juice of the young spadix – Diarrhoea, dysentery, dyspepsia, diabetes, leprosy.

Shell – Strangury.

Kernel – Bronchitis, Skin diseases, Tumours.

Water – Diarrhoea, Dyspepsia, Strangury.

Oil – Bronchitis, cough, diabetes, strangury.

126. *Coffea arabica* L. (Rubiaceae)

Vern.: Kaphi.

Distribution: Planted in gardens and forest nurseries.Hoshangabad.

Botany: A glabrous large shrub, with white flowers and berry fruits.

Flowering & Fruiting: March-December.

Medicinal Parts: Seeds

Chemical constituents: Seed contain 2-Methyl-3-furanthiol, 2-furfurylthiol, prenylmerceptan, β-galactosidase, tocopherols, theobromine, theothylline, trigonelline, chlorogenic acid, caffeine, cafestol, isochlorgenic acid, neochlorogenic acid, 3-feruloylquinic acid, palmitic acid, stearic acid, oleic acid, linoleic acid, cafaryloside, atractyligenin glycoside, stigmasterol, campesterol, sitosterol, cholesterol, cafestol.

Medicinal Properties: Seeds – Antipyretic, bitter, diuretic, stimulant.

Medicinal Uses: Unripe seeds – Fever, gout, migraine.

Ripe roasted seeds – Diarrhoea.

127. *Coix lacryma* – jobi L. (Poaceae)

Vern.: Motida.

Distribution: Gwalior, Hosahangabad, Indore, Mandla.

Botany: Robust, spongy, annual grass, often rooting at lower nodes. Leaves cordate at base, with a glabrous membranous ligule. Racemes consisting solitary basal female spikelet and several male spikelets upwards with glume hardly or narrowly winged.

Flowering & Fruiting: September – November.

Medicinal Parts: Roots, seeds.

Medicinal Properties: Seeds – Bitter, cathartic, depurative, diuretic, sweet, tonic.

Medicinal Uses: Root – Menstrual disorders, straungury.

Seed – Pectoral diseases, pneumonosis.

128. *Coleus aromaticus Benth.* (Lamiaceae)

Syn.: *C. amboinicus* Lour.

Vern.: Patthar chur.

Distribution: Planted in Gardens.

Botany: A small perennial, hairy herb. About 1-3 ft. high with thick and fleshy aromatic leaves; flowers light purple.

Flowering & Fruiting: February-March.

Medicinal Parts: Leaves.

Chemical constituents: Leaves contain Diterpene, Methylene – quinone, couleon, cyclobuta tucin.

Root contains Naphtho Pyrone (Rocetin) and Dialdehyde, Avolatile oil carvacrol is also found.

Medicinal Properties: Leaf – Acrid, anodyne, anthelmintic, appetiser, aromatic, bitter, carminative, constipating, digestive, diuretic, expectorant, stomachic, thrmogenic.

Medicinal Uses: Leaf – Asthma, bronchitis, cholera, colic, cough, diarrhoea, dyspepsia, epilepsy, fever, flatulence, hiccough, otalgia, renal & vesical caliculi.

129. *Colocasia esculenta* (L.) Schott. (Araceae)

Syn.: *C. antiquorum* Schott.

Vern.: Arvi.

Distribution: In moist and shady places of the forests. Balaghat, Mandla, W. nimar.

Botany: Perennial herbs, with tuberous root – stock of various sizes, shape. Leaves cordate or sagittate at the base, entire, purple – blotched. Spadix with an apical barren appendage, shorter than spathes; spathes pale – yellow.

Flowering & Fruiting: July – September.

Medicinal Parts: Leaves, Corms.

Medicinal Properties: Leaf – Rubefacient, stimulant, styptic.

Corm – Anodyne, demulcent, laxative.

Medicinal Uses: Leaf – Internal haemorrhages, otalgia, otorrhoea.

Corm – Haemorrhoids, somatalgia.

130. *Commelina benghalensis* L. (Commelinaceae)

Distribution: Abundant in moist places mostly in alluvial soil, often in wastelands. Balaghat, Damoh, Hosahangabad, Indore, Mandla, Panna, Rewa, Sidhi, Tikamgarh, W. Nimar.

Botany: Diffusely branched, deumbent herbs, with elliptic-ovate leaves. Flowers blue; spathes sessile or subsessile. Capsules 3-gonous; seeds pitted.

Flowering & Fruiting: August-January.

Medicinal Parts: Whole plant.

Chemical Constituents: The flower pigment contains delphinidin diglucoside as the main constituent and P-cumaric acid and awobanol.

Medicinal Properties: Whole plant – Bitter, Demulcent, Emollient, Laxative, Refringent.

Medicinal Uses: Whole plant – Leprosy.

131. *Convolvulus arvensis* L. (Convolvulaceae)

Distribution: Common as weed in fields, along railway tracks & forests roads. Damoh, Indore, Raisen, Satna, Seoni, Sidhi.

Botany: Creeping or twining, perennial herbs, Leaves sagittate-hastate at base, Flowers pale pink,

Flowering & Fruiting: December-February.

Medicinal Parts: Whole plant.

Chemical Constituents: Roots – Cuscohygrine, Coumarins.

Aerial parts – N-alkanes, N-alkanols, α-amyrin & Sterols.

Medicinal Properties: Plant extract – Anticonvulsant, Hypotensive, Mascarinic & Nicotinic.

Medicinal Uses: Leaves – Boils.

132. Convolvulus pluricaulis Choisy. (Convolvulaceae)

Vern.: Shankpushpi.

Distribution: Common in open fields, sandy and rocky places. Chhatarpur, Gwalior, Indore, Mandla, Rewa, Seoni, Sidhi.

Botany: A prostrate or suberect, spreading, hairy, perennial herb with a woody rootstock. Leaves linear-lanceolate; hairy. Flowers white or pale-rose, solitary or in pairs, sessile or subsessile in the leaf axils. Capsule oblong-globose.

Flowering & Fruiting: December-February.

Medicinal Parts: Whole plant, root, leaves, flowers.

Chemical constituents: Plant contains Alkaloid, shankhpushpin, essential oil, methyl pentos, scamonolic acid and palmintic acid, phytosterol, ipuranol, sucrose, reducing sugars, starch and olein.

Medicinal Properties: Anthelmintic, brain tonic, laxative, spasmolytic, stomachic.

Medicinal Uses: Effective on brain and heart, effective on thyroid high secretion; increases memory, positive effect on blood pressure, reduces blood pressure and tension, useful in epilepsy, insomnia & mental diseaes.

133. *Corchorus olitorius* L. (Tiliaceae)

Distribution: Balaghat, Damoh, Dhar, Gwalior, Hosahangabad, E. Nimar, Mandla, Panna, Satna, Shivpuri, Sidhi.

Botany: Annual herbs, with ovate-lanceolate leaves & yellow flowers. Capsules 5-valved, 10 ribbed, with entire beak.

Flowering & Fruiting: September-January.

Medicinal Parts: Leaves, Seeds.

Chemical Constituents: Fibers – Cellulose, Lignin, Nitrogen, Fat, Wax.

Seeds – Crude protein, Crude fibre, Free invert sugar, Minerals; Fixed oil; Fatty acids; Oleic & Linoleic acid, Arachidic acid, Palmitic, Stearic acid; Crystalline Substance-Corchoritin.

Medicinal Properties: Leaves – Acrid, Alexteric, Alterative, Astringent, Bitter, Demulcent, Diuretic, Hot, Tonic.

Seeds – Purgative.

Medicinal Uses: Leaves – Abdominal tumours, Ascites, Pain, Piles, Tumours.

134. *Cordia dichotoma* Forst. f. (Boraginaceae)

Syn.: *C. obliqua* Willd.; C. myxa Roxb. non L.; *C. latifolia* Roxb.

Vern.: Gunda.

Distribution: Common along roadsides, near villages and forest edges. Chhatarpur, Damoh.

Botany: Deciduous trees, with 3 to 5-nerved, elliptic-ovate or obovate leaves longer than broad. Flowers white, in dichotomous cymes. Fruits globose, yellow to red when ripe.

Flowering & Fruiting: March – July.

Medicinal Parts: Bark, leaf, fruits.

Chemical constituents: Bark contains Tannins, allantonin, B-sitosterol, a flavanone rhamnopyranoside.

Fruits contain Mucilage & sugars.

Seed oil contain B-sitosterol.

Seeds contain flavonolglycosides, taxifolin-3 – rhamnoside.

Medicinal Properties: Bark – Acrid, anthelmintic, astringent, bitter, constipating, cooling, depurative.

Leaf – Aphrodisiac.

Fruits – Anthelmintic, aphrodisiac, cooling, depurative, diuretic, emollient, expectorant, febrifuge, purgative, sweet.

Medicinal Uses: Bark – Burning sensation, diarrhoea, dyspepsia, fever, helminthiasis, leprosy, skin diseses.

Leaf – Gonorrhoea.

Fruits – Bronchitis, burning sensation, cough, fever, leprosy, pectoral diseases, ringworm, skin diseases, ulcers.

135. *Coriandrum sativum* L. (Apiaceae)

Vern.: Dhania.

Distribution: Widely cultivated throughout the State. Chhatarpur, Sidhi.

Botany: An annual glaberous herbs, with a strong smell; lower leaves palmatipartite, upper leaves ternately decompound; flowers white, dull purplish in the centre; fruits elliptic-orbicular, glabrous.

Flowering & Fruiting: December-April.

Medicinal Parts: Leaves, Fruits.

Chemical constituents: Leaves – Oxalic acid, calcium, vitamin-C.

Fruits – β-sitosterol, chlorogenic acid, caffic acid, umbelliferone, scopoletin.

Seeds – Quercetin 3-0-caffeyl glucoside, kaempferol 3-glucoside, fatty oil, essential oil.

Medicinal Properties: Leaves – Acrid, anti-inflammatory, aromatic, astringent, styptic.

Fruits – Acrid, anit-inflammtory, appetizer, aromatic, astringent, bitter, carminative, digestive, diuretic, emollient, stimulant, sweet, thermogenic.

Medicinal Uses: Leaves – Hiccough, Inflammations, Jaundice.

Fruits – Bronchitis, burning sensation, cough, diarrhoea, dropsy, dysentery, erysipelas, helminthiasis, scrofula.

136. *Costus speciosus* (Koenig ex Retz.) J.E. Smith (Zingiberaceae)

Syn.: *Banksea speciosa* Koenig.

Vern.: Kevuk kand.

Distribution: Common in shady places along rivers or drains, sometimes in sal forests. Balaghat, Damoh, Hosahangabad, Rewa, Sidhi, W. Nimar.

Botany: An erect, herbaceous plant; leaves subsessile, arranged spirally, oblong or oblanceolate, glabrous above, thinly silky beneath; flowers white, large, many, in terminal dense heads; bracts 2-3 cm long, ovate, brownish-red or bright red; capsule subglobosely 3-gonous; seeds black with white aril.

Flowering & Fruiting: August-May.

Medicinal Parts: Rhizomes.

Chemical constituents: Rhizomes contain Alkaloids, Tigogenin, Diosgenine, Starch, β-Sterol, glucoside.

Leaves contain Alpha-Amyrin stearate, β – Amyrin and Lupeol palmitates.

Medicinal Properties: Rhizome – Acrid, anthelmintic, aphrodisiac, astringent, bitter, cooling, depurative, expectorant, febrifuge, purgative.

Medicinal Uses: Rhizome – Anaemia, asthma, bronchitis, burning sensation, constipation, fever, flatulence, helminthiasis, hiccough, inflammation, leprosy, skin diseases.

137. *Crinum defixum* Ker.-Gwal. (Amaryllidaceae)

Vern.: Vishamandala.

Distribution: In Marshy and muddy places along streams and roadside ditches. Bhopal.

Botany: A perennial herb with avoid bulb; leaves linear, subfleshy, slightly concave, apex acute; scape 30-50 cm. long; bracts spathaceous, membranous; flowers white; fruit sub-globose, beaked; seeds 1-2, rogouse.

Flowering & Fruiting: July-October.

Medicinal Parts: Bulb, Leaves.

Chemical Constituents: *Bulb* – Lycorine.

Medicinal Properties: Bulb – Diaphoretic, Emetic, Emollient, Nauseant.

Medicinal Uses: Bulb – Burns, Carbuncle, Witlow.

138. *Crotalaria retusa* L. (Fabaceae)

Vern.: Ghunghunia.

Distribution: Balaghat, Hoshangabad, Mandla, Satna, Seoni.

Botany: Erect undershrubs; branches striate, appressedly white pubescent; leaves obovate, oblanceolate, and silky pubescent beneath; flowers in 15-30 flowered terminal racemes; pod exserted glabrous, stalked, linear-oblong; seeds 15-30.

Flowering & Fruiting: November-April.

Medicinal Parts: Whole plant.

Medicinal Properties: Plant – Astringent, bitter, emetic, expectorant.

Seed – Depurative, digestive, febrifuge.

Medicinal Uses: Plant – Cardiac disorders, dyspepsia, fever.

Leaf – Diarrhoea, scabies.

Seed – Fever, flatulence, leprosy, skin diseases.

139. *Cryptolepis buchanani* Roem & Schult. (Asclepiadaceae)

Vern.: *Nagbel.*

Distribution: In Sal and mixed deciduous forests. Balaghat, Chhatarpur, Damoh, Hosahangabad, Indore, Mandla, Panna, Rewa, Satna, Seoni,

Botany: Large twiners; flowers Pale Yellow in axillary cymes.

Flowering & Fruiting: May-February.

Medicinal parts: Root, Whole plant

Chemical constituents: *Root contains* – Sarmentogenin, Sarmentocymarin, Buchanin, Cryptanoside C& D.

Leaf & *Stem* – Buchananine, Cryptolepine, Cryptanoside A & B.

Medicinal properties: Aromatic, Anti-dysenteric, Anti-rheumatic, Cooling, Galactagogue.

Medicinal uses: Abdominal diseases, Bodyache, Cholera, Dysentery, Rickets, Bone fractures, Paralysis.

140. *Cucumis sativus* L. (Cucurbitaceae)

Vern.: *Khera.*

Distribution: Cultivated. Bhopal, Chhindwara.

Botany: A scabrous, trailing or climber annual. Leaves 3 to 5 lobed, cordate, very rough. Flowers Yellow; Male ones in axillary clusters and females solitary or fascicled. Fruits ovate or oblong, very variable in size. Seeds numerous, White.

Flowering & Fruiting: May-September.

Medicinal Parts: Fruits, Seeds.

Chemical Constituents: Fruits contain – Erepsin enzyme, Vitamin B & C; proteolytic enzymes, Ascorbic acid, Oxidase, Succinic & Malic dehydrogenases, rutin, Palmitic, Stearic, Oleic acids, their Glycerides. Sterols & Squalene.

Seeds contain – Sitosterol, Cucurbitaside.

Leaf contains – Ferrodoxin, free Cucurbitasides B & C.

Medicinal Properties: Fruits – Diuretic, Haemostatic, Refrigerant, Sweet, Tonic.

Seeds – Antipyretic, Astringent, Diuretic, Nurtitive, Purgative, Refrigerant, Sweet, Tonic.

Medicinal Uses: Fruits – Bronchitis, Burning sensation, Fever, General debility, Haemorrhages, Insomnia, Jaundice, Strangury.

Seeds – Burning sensation, Constipation, General debility, Intermittent fever, Renal calculus, Strangury.

141. *Cucumis trigonus* Roxb. (Cucurbitaceae)

Syn.: *C. callosus* Cogn.

Vern.: *Ban kachria.*

Distribution: Balaghat, Betul, Satna, Sidhi.

Botany: Perennial tendrillar herbs, with palmately five-lobed leaves and Yellow unisexual flowers.

Flowering & Fruiting: **June-October.**

Medicinal Parts: Roots, Fruits, Seeds.

Chemical Constituents: Seed oil – Linoleic acid, Oleic acid, Stearic aicd, Arachidic acid, Behenic acid, Myristic acid, Palmitic acid.

Fruits – Cucurbitacin-B, D & Q, Chondrillasterol, β-Stiosterol.

Medicinal Properties: Fruit pulp – Acrid, Anthelmintic, Appetiser, Bitter, Expectorant, Febrifuge, Purgative, Stomachic, Thermogenic.

Seeds – Astringent, Cooling.

Medicinal Uses: Fruit pulp – Anaemia, Bronchitis, Constipation.

Cough, Diabetes, Fever, Flatulence, Helminthiasis, Jaundice, Leprosy.

142. *Cucurbita maxima* Duchesne ex Lam. (Cucurbitaceae)

Vern.: Kaddu.

Distribution: Widley cultivated. Bhopal.

Botany: Leaves circular to reniform in outline. Flowers Yellow. Peduncle short, spongy.

Flowering & Fruiting: March-August.

Medicinal Parts: Fruits, Seeds.

Chemical Constituents: Leaves – Calcium, Magnesium, Iron, Zinc, Copper.

Seed – Euglobulin, □ -Spinasterol, Linoleic acid, Oleic acid, Linolenic acid, β-tocopherol, β-Sitosterol, Cholesterol, Ergosterol, Campesterol, Codisterol, Clerosterol, Isofucosterol.

Medicinal Properties: Fruits – Diuretic, Emollient, Refrigerant, Sedative, Sweet, Tonic.

Seeds – Anthelmintic, Diuretic, Nervine tonic.

Medicinal Uses: Fruits – Boils, Burns, Inflammation, Migraine, Neuralgia, Scalds.

Seeds – Nervous debility, Strangury.

143. *Cucurbita pepo* L. (Cucurbitaceae)

Vern.: Kela.

Distribution: Cultivated.

Botany: An annual climbing herb.

Flowering & Fruiting: Rainy season

Medicinal parts: Seeds & Leaves.

Chemical constituents: *Flower contains* – Flavonol glycosides.

Seeds contain – Cucurbitine, Gibberellins & Kaurenolide, Phytosterarin, Salicylic acid.

Medicinal properties: *Fruit* – Cooling, Astringent, Appetizer.

Leaves – Haematic, Analgesic.

Seeds – Anthelmintic.

Medicinal uses: *Fruit* – Leprosy.

144. *Cuminum cyminum* L. (Apiaceae)

Vern.: Jeera.

Distribution: Widely cultivated throughout the State.

Botany: Glabrous, annual herbs, with Rose-coloured flowers. Fruits setulose, narrowly oblong.

Flowering & Fruiting: January-March.

Medicinal Parts: Fruits.

Chemical constituents: Fruits contain Apigenin-7-O-glucoside, Luteolin-7-O-glucoside, essential oil.

Medicinal Properties: Fruits – Acrid, anaphrodisiac, anodyne, anthelmintic, anti-inflammatory, astringent, carminative, constipating, cooling, depurative, digestive, febrifuge, stomachic, sweet.

Medicinal Uses: Fruits – Asthma, colic, dyspepsia, fever, flatulence, helminthiasis, inflammations, leprosy, leucorrhoea, skin diseases, strangury, vomiting.

145. *Curculigo orchioides* Gaertn. (Amaryllidaceae)

Syn.: *C. malabarica* Wight.; *C. brevifolia* Dryland.; *Hypoxis orchiodies* Kurz.; *Carcinia malabarica* Desv.

Vern.: Kali musli.

Distribution: In Sal forests. Throughout the State.

Botany: Perennial herbs, with thick tuberous root – stock and without aerial stem. Leaves plicate, linear – lanceolate. Flowers yellow 2 to 3 together on axillary peduncles hidden among leaf-sheaths.

Flowering & Fruiting: June – September.

Medicinal Parts: Root-stock.

Chemical constituents: Root stock contains Steroids–Terpenoids, Phenolics, Tannins, Glycosides, Starch, Iron, Calcium, Magnessium, Sodium, Stranthium, Resin, gum, Calcium Oxalate.

Medicinal Properties: Alterative, antipyretic, aphrodisiac, appetiser, bitter, carminative cooling, depurative, diuretic, emollient, sweet.

Medicinal Uses: Asthma, bronchitis, colic, diarrhoea, dyspepsia, gonorrhoea, jaundice, leucorrhoea, lumbago, pruritus, skin diseases, vomiting, wounds.

146. *Curcuma amada* Roxb. (Zingiberaceae)

Vern.: Ambi haldi.

Distribution: In moist places between rock boulders or near streams. Balaghat, Bhopal, Hosahangabad, Indore, Satna, W. Nimar.

Botany: Stemless herbs, with ellipsoid, sessile tubers with the smell of unripe mango. Leaves oblong-lanceolate. Flowers white or pale-yellow, in spikes appearing with the leaves.

Flowering & Fruiting: July-October.

Medicinal Parts: Rhizomes.

Chemical constituents: Rhizome contains Essential oil (Pinene, ocimene, safrol), Resin, sugars, Gum, Starch, Albuminoids, crude fibres, organic acid, Phytosterol, zulenogeine, Camphor, curcumene, Turmerone, Volatile oil.

Medicinal Properties: Anti – inflammatiory, anti-pyretic. aphrodisiac, appetizer, aromatic, bitter, carminative, cooling, digestive, diuretic, expectorant, febrifuge, laxative, sour, stomachic, sweet.

Medicinal Uses: Bronchitis, colic, constipation, cough, dyspepsia, fever, flatuance, gout, hiccough, inflammation, pruritus, skin diseases, ulcers, wounds.

147. *Curuma longa* L. (Zingiberaceae)

Syn.: *C. domestica* Valeton.; *C. rotunda* L.; *Amomum curcuma* Jacq.

Vern.: Haldi.

Distribution: Cultivated. Bhopal, Indore and other districts of the State.

Botany: Rhizomes large, yielding turmeric. Leaves oblong-elliptic. Flowers pale-yellow.

Flowering-Fruiting: September-November.

Medicinal Parts: Rhizomes.

Chemical constituents: Rhizome contains Active curcumin compound; essential oils-Termeric oil, Terpenoids, A volatile oil: it contains curcumen (Terpene), starch, Albumnoids.

Medicinal Properties: Rhizome – Acrid, alterative, anodyne, anthelmintic, anti – inflammatory, antiperiodic, antiseptic, appetiser, bitter, carminative, depurative, diuretic, emollient, expectorant, febrifuge, laxative, stimulant, stomachic, thermogenic.

Medicinal Uses: Rhizome – Amenorrhoea, anaemia, asthma, bronchitis, colic, conjunctivitis, constipation, cough, diabetes, dropsy, dyspepsia, elephantiasis, fever, flatulence, gonorrohoea, haemorrhoges, helminthiasis, hiccough, hysteria, inflammation, jaundice, leprosy, pruritus, ringworm, skin diseases, ulcers, wounds.

Ayurvedic preperations: *Haridrakhand, Chandrabrabha vati, Laghu-vis-garva-taila, Pippalasav, Punarnava mandur, Sudarshan churna, Vidangadi lep, Geriforte, Majnoon-e-falsfa, Habs harkachur, Dhatrinisa, Haridradi kwath, Haridradi churna, Haridradi ghirit, Haridradi yoga, Haridradi taila.*

148. *Curcuma zedoaria* Rosc. (Zingiberaceae)

Vern.: **Banhaldi.**

***Distribution*:** In forest clearings. Chhattarpur, Sidhi.

Botany: An erect herb with rhizome.

Flowering & Fruiting: May-August.

Medicinal parts: Tuber & Leaves.

Chemical constituents: An essential oil, a bitter soft resin, Organic acids, Gum, Starch, Resin, Sugar, Curcumin arabins, Albuminoids, Crude fibre & Ash.

Medicinal properties: *Rhizome* – Carminative, Stomachic, Aromatic, Cooling, Diuretic.

Medicinal uses: *Rhizome* – Dyspepsia, Flatulence.

149. *Cuscuta reflexa* Roxb. (Convolvulaceae)

Vern.: Amarbel.

Distribution: Common on a variety of hosts. Balaghat, Betul, Chhatarpur, Damoh, Hosahangabad, Indore, Narsimhapur, Seoni, Shahdol, Sidhi.

Botany: Leafless stem – parasites, with greenish – yellow stem. Flowers whitish, solitary or in clusters arranged in racemes; style 1; stigmas 2.

Flowering & Fruiting: November-March.

Medicinal Parts: Whole plant.

Chemical constituents: Plant contains Alkaloids, proteins, flavonoids, resin, tannins, glycosides, carbohydrates, cuscutin, cuscatalin, reducing sugars.

Seeds contain Fixed oil, Amarbelin (Pigment.), bitter resin, aluminum, calcium, sodium, potassium & reducing sugars.

Medicinal Properties: Whole plant – Anthelmintic, aphrodisiac, astringent, bitter, carminative, diuretic, expectorant, sweet.

Medicinal Uses: Whole plant – Bronchitis, cough, fever, Jaundice, paralysis.

150. *Cyamopsis tetragonoloba* (L.) Taub. (Fabaceae)

Syn.: *C. psoralioides* (L.) DC.

Vern.: **Guar.**

Distribution: Raisen, Sagar.

Botany: An erect robust annual. Leaflets 3, ovate-elliptic, Coarsely toothed. Flowers Pinkish-purple, in axillary racemes. Pods erect. Seeds dull, Compressed.

Flowering & Fruiting: *April-December.*

Medicinal Parts: Pods.

Chemical Constituents: Pods – Methionine, Globulin, Gallic acid, Caffic acid, Chlorogenic acid, Ellagic acid, Gallotannins, Kaempferol.

Seeds – Gallic acid, Galactose, Saponins.

Medicinal Properties: Pods – Appetiser, Cooling, Digestive, Laxative, Sweet, Tonic.

Medicinal Uses – Pods – Agalactia, Anorexia, Constipation, Dyspepsia.

151. *Cymbopogon jwarancusa Schult* (Poaceae)

Syn.: *Andropogon jwarancusa* Jones.

Distribution: Chhattarpur.

Botany: A highly aromatic Grass.

Flowering & Fruiting: August-November.

Medicinal parts: Stem, Leaves, Flowers & Oil.

Chemical constituents: Essential oil.

Medicinal properties: Stimulant, Blood Purifier, Sudorific.

Medicinal uses: Gout, Fever, Chronic rheumatism, Cholera, dyspepsia.

152. *Cymbopogon martinii* (Roxb.) Wats. (Poaceae)

Syn.: *Andropogon martini* Roxb.

Vern.: Rosha.

Distribution: Common. Gwalior, Hosahangabad, Sidhi.

Botany: Tufed, sweet-scented perennial grass. Leaves amplexicaul at base. Panicles large, decompound; sessile spikeletes awned and lower glume have a deep median groove.

Flowering & Fruiting: September– February.

Medicinal Parts: Whole plant.

Chemical constituents: Leaves contain Geranium oil.

Medicinal Properties: Acrid, appetiser, bitter, carminative, depurative, digestive, diuretic, febrifuge.

Medicinal Uses: Bronchitis, catarrh, cardiac debility, colic, cough, dyspepsia, fever, helminthiasis, leprosy, skin diseases.

153. *Cynodon dactylon* (L.) Pers. (Poaceae)

Syn.: *Panicum dactylon* L.

Vern.: Dub.

Distribution: Common throughout the State.

Botany: Creeping, rhizomatous grass. Ligule a ring of hairs. Spikes 2-6, up to 6 cm long; lemmas longer than upper glumes, simple hairy. Grains turgid.

Flowering & Fruiting: Most throughout the year.

Medicinal Parts: Whole plant.

Chemical constituents: Plant contains β – Sitosterol, β – Carotene, Vitamin – C, Carotenes, Glucosides, Palmitic acid, Triterpenoid, Arundoin, Friedelene, Alkaloids – Erganovin, Erganovinine, ferulic, mono & oligo, Saccharide – glucose, fructose, sucrose, starch & lignin.

Medicinal Properties: Astringent, cooling, constipating, depurative, diuretic, sweet.

Medicinal Uses: Abortion, Burning sensation, conjunctivitis, diarrhea, dysentery, general debility, haemorrhages, haemorrhoids, leprosy, skin diseases, vomiting, wounds.

Ayurvedic preperations: *Durvadi kwath, Durvadi ghirit, Durvadi taila, Durvadi yoga.*

154. *Cyperus esculentus* L. (Cyperaceae)

Distribution: Satna.

Botany: An erect perennial glabrous grass-like sedge, upto 50 cm. in height with slender subterranean stolons ending in ovoid or cylindrical brown edible tubers; leaf blades shorter than stems; spikelets in umbellate peduncled spikes, flowering glumes boat shaped; Fruit abovoid-ellipsoid nut with 3-sharp angles.

Flowering & Fruiting: August.

Medicinal Parts: Tubers.

Taste: Bitter, Astringent.

Medicinal Properties: Tubers – Acrid, anthelmintic, aphrodisiac, astringent, bitter, carminative, cooling, digestive, febrifuge.

Medicinal Uses: Tubers – Colic, diarrhoea, dysentery, dyspepsia, flatulence, helminthiasis, leprosy.

155. *Cyperus rotundus* L. (Cyperaceae)

Syn.: *C. hexastachys* Rottb.

Vern.: Motha.

Distribution: In open moist places. Betul, Damoh, Hosahangabad, Indore, Mandla, Rewa, Satna, Seoni, Sidhi, Ujjain, W. Nimar.

Botany: Sedges, with slender stolons terminated by ellipsoid or globose-ovoid tubers. Leaves not exceeding the clum. Spikes born in terminal corymbs; spikelets dusty brown, 8 to 28 flowered; rachilla winged. Achenes triquetrous, brown.

Flowering & Fruiting: July-April.

Medicinal Parts: Tubers.

Chemical constituents: Rhizome contains Essential oil which contains; Penene, cineol; fatty oil which contains; Linoleic acid, oleic acid, Myristic acid, Stearic; Glycerol, Glycoside cyperene, cyperenone, sugars, gum, carbohydrates.

Leaves contain Luteolin.

Medicinal Properties: Acrid, anthelmintic, anti – inflammatory, antiperiodic astringent, bitter, carminative, constipating, cooling, depurative, digestive, diuretic, expectorant, febrifuge, intellect promoting, stomachic.

Medicinal Uses: Amenorrhoea, amentia, bronchitis, colic, cough, diarrhoea, dysmenorrhoea, dysentery, dyspepsia, epilepsy, fever, flatulence, general calculi, general debility, inflammation, leprosy, pruritus, scabies, skin diseases, ulcer, verminosis, vomiting, wounds.

Ayurvedic preperations: *Mustakadi kwath, Mustakarisht, Mustadi churna, Mustadileh, Shandgapaniya, Jawarish jalinush, Sadangodaka, Musta taila, Majun-murrawal-ul-anwah, Balasanjivani churna.*

156. *Dalbergia sissoo Roxb.* (Fabaceae)

Vern.: Shisham.

Distribution: On rosdsides and in mixed forests. Balaghat, Bhopal, Chhatarpur, Damoh, Gwalior, Hosahangabad, Indore, Jabalpur, Mandla, Ratlam, Rewa, Sagar, Satna, Seoni, Shahdol, Shivpuri, Sidhi.

Botany: Deciduous tree, with 3-5, acuminate leaflets. Flowers yellowish, in axillary panicles. Stamens 9.

Flowering & Fruiting: February-June.

Medicinal Parts: Root, Leaves, bark, heartwood.

Chemical constituents: Plant contains Dalbergin, dalbergenone, methyl dalbergin, nordalbergin, Isodalbergin, dalbergiphenol, tecterigenin, Isocaviudin, glucoside, Isocaviunin, essential oil which contain: Bisboline & nerolidol.

Medicinal Properties: Root – Astringent, constipating.

Leaf – Anthelmintic, Bitter, constipating, digestive, diuretic, styptic.

Bark/heartwood – Abortifacient, acrid, anthelmintic, anti – inflammatory, antipyretic, appetiser, astringent, bitter, depurative, thermogenic.

Medicinal Uses: Root – Diarrhoea, dysentery.

Leaf – Buring sensation, colic, diarrhoea, dysentery, dyspepsia, gonorrhoea, menorrhoea, vomiting.

Bark/heartwood – Amenorrhoea, dysmenorrhoea, burning sensation, bronchitis, dysentery, dyspepsia, fever, gout, hiccough, inflammation, leprosy, leucoderma, scabies, sciatica, skin diseases, syphilis, ulcers, vomiting.

157. *Datura metal* L. (Solanaceae)

Syn.: *D. alba* Nees.; *D. fastuosa* L. var. *alba* C. B. Clarke.

Vern.: Kala dhatura.

Distribution: In wastelands and near drains throughout the State. Bhopal, Chhatarpur, Damoh, Indore.

Botany: A shrubby herb with dichotomous branches. Leaves ovate, sinuate, unequal sided at the base. Flowers purple or white, large, funnel shaped. Persistent calyx. Capsule globose, prickled.

Flowering & Fruiting: July – March.

Medicinal Parts: Whole plant.

Chemical constituents: Seeds & flowers contain Daturadiol, Daturalone, fastusine, β – sitosterol, hyosine, hyoscyamine, atropine, fastudin, fastunine, fastusic acid, fastusidine, fastusinine, daluranlone, citrostadienol, obtusifoliol, allantoin, caproic, β – linolic, linoleic acid, oteic acid, datumetine, datumetelin, vitamin – C, chlorogenic acid, fixed oil.

Medicinal Properties: Root – Anodyne, antispasmodic, narcotic.

Seed – Antispasmodic, aphrodisiac, narcotic.

Medicinal Uses: Leaf – Asthma, dandruff, epilepsy, lumbago, mumps, neuralgia, otalgia, sciatica, swellings.

Seed – Asthma, otalgia, skin diseases.

Ayurvedic preperations: *Kanakasav, Dhaturdhum, Martyuanjaya ras, Dattuphal vasma, Dhatturadi pralep, Laghu-vis-garva taila, Dugdha vati, Jawrankush, Haba shafa, Roghan dhatura, Kanakasundara, Laksmivilas rasa, Susekhara, Mahajvarankusa.*

158. *Datura stramonium* L. (Solanaceae)

Syn.: *D. tatula* L.

Vern.: Safed dhatura.

Distribution: In Wastelands throughout the State.

Botany: An erect, annual herb; leaves sinuate, lanceolate, acute at apex, narrowed at base, membranous, sparsely pubescent; flowers white; capsule erect, oblong, densely covered with long prickles.

Flowering & Fruiting: November-March.

Medicinal Parts: Leaves, flowers, fruits, seeds.

Chemical constituents: Plant contains Hysocyamine, hyoscine, atropine, scopalamine, tigloidin, tropine, linoleic acid, legnoceric, myristic, palmitic acid, oleic, stearic acid, β – carbolines, compesterol, β-sitosterol, stigmasterol, β – amyrin, daturadiol, gramisterol, chlorogenic acid, datugen, datugenine, daturalactone, stramonolide.

Medicinal Properties: Seed – Acrid, alexiteric, anthelmintic, bitter, emetic, febrifuge, heating, sharp taste, tonic.

Medicinal Uses: Seed – Biliousness, itching, jaundice, leucoderma, piles, skin diseases, ulcer.

Ayurvedic preperations: *Kanakasava, Dhaturdhum, Martyuanjaya ras, Jawarankush.*

159. *Delonix regia* (Boj. ex Hook.) Rafin. (FABACEAE)

Syn.: *Poinciana regia* Boj. ex Hook.

Vern.: Gulmohar.

Distribution: Planted as an ornamental tree. Bhopal, Jabalpur, Rewa, Sagar, Shivpuri.

Botany: Trees, with Yellow flowers & pod fruits. *Flowering & Fruiting:* April-August.

Medicinal parts: All parts.

Chemical constituents: Bark contains – Leucocyanidin, Lupeol.

Medicinal properties: *Bark* – Febrifuge, Antiperiodic.

Medicinal uses: *Plant* – Rheumatism, Flatulence.

Leaf – Leucorrhoea.

160. *Dendrocalamus strictus* (Roxb.) Nees. (Poaceae)

Vern.: *Bans.*

Distribution: Balaghat, Hosahangabad, Indore, Jabalpur, Rewa, Seoni.

Botany: Arborescent or shrubby bamboos, with smooth culms. Leaves with a petiole like base & articulated with sheath. Spikelets in dense clusters along the panicle branches. Rachilla internodes very short, not articulating. Fruits subglobose.

Flowering & Fruiting: *November-April.*

Medicinal Parts: Clum.

Chemical Constituents: Silicious matter.

***Medicinal Properties*:** Clum – Tonic, Astringent, Cooling.

161. *Desmodium gangeticum* (L.) DC. (Fabaceae)

Syn.: *D. collinum* Roxb.; *Hedysarum gangetium* L.

Vern.: Sarivan.

Description: In Sal and Teak forests. Balaghat, Bhopal, Chhindwara, Damoh, Gwalior, Hosahangabad, Indore, Jabalpur, Mandla, Narsimhapur, Panna, Raisen, Rewa, Satna, Seoni, Shahdol, Shivpuri, Sidhi, Tikamgarh.

Botany: Undershrub; leaflets ovate-oblong; flowers purple-pink, clustered in racemes; pods linear, flat, falcate.

Flowering & Fruiting: July-April.

Medicinal Parts: Roots.

Chemical constituents: Plant contains Pterocarpanoids, Gangetin, gengetinin, Desmodin, alkaloids, Isoflavonoid, Phytoalexins, Isoflavones, genestein.

Medicinal Properties: Root – Anti-Inflammatory, aphrodisiac, bitter, carminative, constipating, diuretic, expectorant, febrifuge, sweet, thermogenic.

Medicinal Uses: Root – Asthma, debility, bronchitis, cough, dysentery, dyspepsia, fever, gout, haemorrhoids, inflammation.

162. *Desmodium triflorum* (L.) DC. (Fabaceae)

Distribution: Along sreams. Balaghat, Bhind, Bhopal, Chhindwara, Damoh, Hosahangabad, Indore, Mandla, Morena, Narsimhpur, Panna, Raisen, Rewa, Sagar, Shivpuri, Sidhi.

Botany: A much branched, gregarious, wiry, trailing herbs, rooting at the nodes. Leaflets obovate, obcordate. Flowers Pink or Purplish, 1-3 in the axils of leaves. Pods curved, flat, indented on the lower edge, 3 to 5 jointed. Seeds Brown.

Flowering & Fruiting: September-April.

Medicinal Parts: Whole plant.

Chemical Constituents: Root / Leave / Stem – Trigonelline, Stachydrine, Betaine, Choline, Indole-3-acetic acid, Tyramine, Hypaphorin, 4β-Phenylethylamines.

Medicinal Properties: The plant is Acrid, Cooling, Expectorant, Sweet.

Medicinal Uses: The plant is useful in Bronchitis, Burning sensation, Cough, Dysentery, Erysipelas, Flatulence, Pruritus, Wounds.

163. *Dichrostachys cinerea* (L.) W. & A. (Mimosaceae)

Syn.: *Mimosa cinerea* L.

Vern.: Bilatri, Sagun Kanti.

Distribution Distribution: In dry deciduous forests. Damoh, Hosahangabad, Seoni, Shahdol, Shivpuri.

Botany: Polygamous shrubs, with branches ending into spines. Pinnae 4-14 pairs and leaflets 12-24 pairs. Spikes yellow in upper half and rose-pink or purple below. Pods twisted torulose.

Flowering & Fruiting: June-January.

Medicinal Parts: Root.

Medicinal Properties: Acrid, anodyne, anti-inflammatory, astringent, bitter, constipating, digestive, diuretic, thermogenic.

Medicinal Uses: Diarrhoea, dyspepsia, elephantiasis, inflammation.

164. *Digera arvensis* (L.) DC. (Amaranthaceae)

Syn.: *D. muricata* (L.) Mart.

Vern.: Lat mahuria.

Distribution: In dry sandy places, along roadsides and as garden weed. Damoh, Dhar, Hosahangabad, Indore, Satna.

Botany: Erect or decumbent herbs; leaves ovate-lanceolate to oblong, entire or crenulate, acute, obtuse or abruptly acuminate. Flowers pink, in axillary, 10-30 cm. long racems; bacts oblong, persistent, 2 outer tepals 3-4 mm long, 5-7 nerved; 3 inner tepals shorter, narrow 1-2 nerved; fruits ovate-globose, with 2-fid apex, 1-seeded.

Flowering & Fruiting: July-April.

Medicinal Parts: Flowers and Seeds.

Chemical Constituents: Leaves – The Protein, Fat, Total mineral, Crude fibre, Carbohydrates. Calcium, Iron, Ascorbic acid and β-carotene.

Medicinal Properties: Acrid, Astringent,Cooling, Laxative, Sweetish.

Medicinal Uses: Flower / Seeds – Urinary discharges.

165. *Dioscorea bulbifera* L. (Dioscoreaceae)

Syn.: *D. versicolor* Buch. Ham.; *D. pulchella* Roxb.; *D. sativa* Thunb.; *C. crispata* Roxb.

Vern.: Varahi kand.

Distribution: Common in mixed and Sal forests. Balaghat, Sidhi.

Botany: Slender, glabrous, twining herbs with globose to pyriform tubers; leaves broadly ovate, glabrous, acuminate, with a deeply cordate base; male flowers in raceme like, axillary spikes, greenish-purple; female flowers sessile, axillary or in fascicled spikes, purple; capsule quadrately oblong.

Flowering & Fruiting: November-January.

Medicinal Parts: Tubers, fruits.

Chemical constituents: Plant contains A poisonous glucoside (Dioscorine).

Tuber contains Starch, Vitamin – B, Calcium oxalate, Protein, Iron.

Medicinal Properties: Tuber – Alternative, anthelmintic, appetiser, aphrodisiac, bitter, fattening, pungent, stomachic, tonic.

Medicinal Uses: Tuber – Biliousness, bronchitis, dyspepsia, leucoderma, piles, strangury, tumors, urinary discharges.

166. *Diospyros malabarica* (Desr.) Kostel. (Ebenaceae)

Vern.: Kala tandu.

Syn.: *D. peregrina auct.* non (Gaertn.) Guerke.; *D. embryopteris* Pers.

Distribution: In mixed forests. Chhatarpur.

Botany: A middle sized evergreen trees; leaves reddish; flowers cream coloured; fruits yellow berries.

Flowering & Fruiting: April-following April.

Medicinal Parts: Bark, Leaves, Flowers, Fruits.

Taste: Astringent.

Chemical constituents: Fruit pulp contains β-sitosterol & its glucoside, Betulin, a monohydroxy triterpene ketone, betulinic acid, gallic acid, hexacosane, hexacosanol oleanolic acid, myricyl alcohol from leaves. Saponin from Bark.

Medicinal Properties: Bark – Acrid, anti-inflammatory, astringent, constipating, cooling, depurative, febrifuge.

Leaves – Carminative, diuretic, laxative, styptic.

Flowers – Aphrodisiac, diuretic.

Fruits – Acrid, bitter, carminative, cooling, digestive.

Medicinal Uses: Bark – Burning sensation, burns, diabetes, dysentery, dyspepsia, fever, inflammations, leprosy, pruritus, skin diseases, spermatorrhoea.

Leaves – Burns, dyspepsia, flatulence, scabies, strangury, wounds,

Flowers – Anaemia, leucorrhoea, scabies.

167. *Diospyros melanoxylon Roxb.* (Ebenaceae)

Syn.: *D. tupru* Buch.-Ham.

Vern.: *Biripatta.*

Distribution: Common in dry deciduous forests and roadsides. Balaghat, Damoh, Hosahangabad, Indore, Jabalpur, Mandla, Morena, Panna, Satna, Seoni, Shivpuri, Sidhi.

Botany: Deciduous trees, with sub-opposite & alternate, ovate-elliptic leaves. Male flowers in drooping cymes; anthers awnless. Fruits globose or ovoid.

Flowering & Fruiting: March-August.

Medicinal Parts: Bark, Leaves, Fruits.

Chemical Constituents: Bark – Tannins.

Medicinal Properties: Leaves – Diuretic, Carminative, Laxative, Styptic.

Bark – Astringent.

Fruits – Carminative, Laxative, Diuretic.

Medicinal Uses: Dried flowers – Urinary, Skin & Blood diseases.

Bark – Diarrhoea, Dyspepsia.

168. *Diplocyclos palmatus* (L.) Jeffrey (Cucurbitaceae)

Syn.: *Bryonia laciniosa* L.; *B. palmata* L.; *Bryonopsis laciniosa* (L.) Naud.

Vern.: Ban-kakra.

Distribution: Hoshangabad, Morena, Sidhi.

Botany: Climbers, with 2-fid tendrils and 5-lobed leaves. Flowers greenish – yellow. Fruits globose, 6-striped; seeds with raised projections on both faces.

Flowering & Fruiting: December-January.

Medicinal Parts: Whole plant.

Chemical constituents: *Seed oil* Punicic acid.

Medicinal Properties: Acrid, alterant, anti – Inflammatory, depurative, thermogenic.

Medicinal Uses: Cough, flatulences, general debility, inflammation, skin diseases.

169. *Dolichandrone falcata Seem.* (Bignoniaceae)

Syn.: *Spathodea falcata* DC.

Distribution: In dry places, sometimes in black soil. Chhatarpur, Hosahangabad, Indore, Shivpuri, W. Nimar.

Botany – Small trees with Bluish grey bark and White flowers; fruits long up to 50 cm., curved, flattened; seeds winged.

Flowering & Fruiting: *March-June.*

Medicinal parts: Aerial parts

Chemical constituents: *Leaves contain* – Luteolin, Chrysin, Chrysin-7-rutinoside.

Medicinal properties: Abortifacient, Carminative, Diuretic.

Medicinal uses: Cough, Diabetes, Jaundice, Intestinal worms, Leprosy, Piles, Urinary calculi & Ulcers.

170. *Dregea volubilis* (L.f.) Benth. ex Hook. f. (Asclepiadaceae)

Syn.: *Wattakaka volubilis* (L.f.) Stapf.; *Marsdenia volubilis* Cooke.

Vern.: Nakchhikni.

Distribution: Climber on bushes along roads & in scrub forests. Gwalior, Hosahangabad, Shivpuri.

Botany: Perennial woody twinners with watery sap; flowers in umbellate cymes, Pale green; follicle long green. *Flowering & Fruiting:* March-November.

Medicinal parts: Roots, Tender Shoots, Leaves.

Chemical constituents: Glucoside; Dregein, An alkaloid.

Stem/Leaves contain: Taraxerol, Triterpenoid, Kaempferol, Saponin.

Seed contains: A number of Pregnane Glycosides.

Medicinal properties: Leaves & tender Stalks – Emetic, Purgative.

Plant – Expectorant.

Medicinal uses: Used in Eye diseases & Snake bite.

171. ***Drypetes roxburghii*** **(Wall.) Hurus. (Euphorbiaceae)**

Syn.: *Putranjiva roxburghii* Wall.

Vern.: Putranjiva.

Distribution: Commonly planted as avenue tree, sometimes in wastelands. Chhatarpur, Damoh, Indore, Rewa, Seoni.

Botany: Evergreen trees, with pendent branches and elliptic-oblong leaves. Flowers dioecious, apetalous, fascicled in axillary clusters. Fruits ellipsoid, hard.

Flowering & Fruiting: March –February.

Medicinal Parts: Leaf, seed.

Chemical constituents: Fruit contains Manitol, Saponins.

Seed contains Arachidic acid, Linoleic acid, Palmitic acid, Stearic acid, Glucochlearin, Glucocleomin, Gluco Putranjivin, Putranjivoside, Putranoside, β – Sitosterol.

Leaves contain Stigmasterol.

Bark contains Carboxylic acid, Putric acid, Putranjivic acid.

Medicinal Properties: Leaf – Procreant, refrigerant.

Seeds – Acrid, anti-inflammatory, aphrodisiac, diuretic, procreant, laxative, sweet.

Medicinal Uses: Leaf – Catarrh, fever, sterility.

Seed – Azoospermia, burning sensation, constipation, elephantiasis, inflammation, sterility.

172. ***Echinops echinatus Roxb.*** **(Asteraceae)**

Vern.: Ontkata.

Distribution: Common in dry wasteland, ridges and boundaries of cultivated fields. In dry deciduous forests. Damoh, Shivpuri, Sidhi.

Botany: Thistle-like annual herbs, with pinnatifid, spinescent leaves white woolly beneath. Heads white, blue or yellow, 1 – flowered clustered in spinos globose compound heads.

Flowering & Fruiting: October – March.

Medicinal Parts: Root.

Chemical constituents: Plant Hentriacontane, hentriacontanol, B-amyrin and leupeol.

Seeds Echinopsine.

Medicinal Properties: Abortifacient, aphrodisiac, bitter, cooling, pungent, sweet.

Medicinal Uses: Biliousness, diseases of the heart, gleet, strangury, urinary discharges.

173. *Eclipta prostrata* (L.) L. (Asteraceae)

Syn.: *E. alba* Hassak.

Vern.: Bhangra.

Distribution: Common in moist waste places. Damoh.

Botany: Erect or prostate herbs, often rooting at the lower nodes. Leaves elliptic-oblong. Heads white, solitary.

Flowering & Fruiting: Throughout the year.

Medicinal Parts: Whole plant.

Chemical constituents: Plant contains Glucosides – phytosterol – A, β – Amyrin, wedelolactone, luteoline, triterpenic acid, fatty acids, palmitic acid, stearic acid, oleic acid, linolic acid, nicotine, ampliptene alkaloid, ecliptine & resin.

Medicinal Properties: Acrid, alterative, anodyne, anthelmintic, anti – inflammatory, aphrodisiac, bitter, carminative, depurative, digestive, diuretic, febrifuge, thermogenic.

Medicinal Uses: Debility, elephantiasis, fever, hypertension, inflammation, jaundice, leprosy, otalgia, pruritus, skin diseases, ulcers, wounds.

Ayurvedic preperations: *Bhringraj taila, Shadbindu taila, Bhringraj ghrit, Bhringrajadi churna.*

174. *Embelia tsjeriam – cottam* (R. & S.) DC. (Myrsinaceae)

Syn.: *E. robusta C. B.* Clarke

Vern.: Bayabirang.

Distribution: Common in Sal and mixed forests. Balaghat, Chhindwara, Hosahangabad, Panna, Sagar, Satna, Seoni, Sidhi.

Botany: Deciduous shrubs or small trees; leaves broadly elliptic; fruit red, globose. *Flowering & Fruiting:* July – January.

Medicinal Parts: Fruits.

Medicinal Properties: Fruits – Carminative, anthelmintic, antispasmodic.

Medicinal Uses: Fruits – Piles, tuberculous glands.

175. *Emilia sonchifolia* (L.) DC. (Asteraceae)

Distribution: Common along river banks; on sandy soil. Damoh, Hosahangabad, Sidhi.

Botany: Variously branched procumbent annual herbs, with Purple flowers and Achene fruits.

Flowering & Fruiting: February-March.

Medicinal Parts: Whole plant.

Chemical Constituents: Aerial parts contain Kaempferol-3β-D Galactoside, Quercitrin, Rutin, Quercetin, Ursolic acid, N-hexacosanol, Triacontane, Hyperoside, Pyrrolizidine, Alkaloids as Senkirkine, Cloronine, β-Sitosterol, Ralmitic acid, Stigmasterol.

Medicinal Properties: The plant is Antipyretic, Astringent, Opthalmic, Sudorific, Sweet, Thermogenic, and Vulnerary.

Medicinal Uses: The plant is useful in Asthma, Cuts, Diarrohea, Intermittent Fever, Ophthalmia, Otalgia, wounds.

176. *Enicostemma littorale Blume.* (Gentianaceae)

Syn.: *E. axillare* (Lam.) Raynal.

Vern.: Nau.

Distribution: Along Ponds. Chhatarpur, Damoh.

Botany: Perennial herb, with 4-angular branches and linear-lanceolate leaves. Flowers white, in axillary clusters. Capsules 1-celled, 2-valved.

Flowering & Fruiting: July – October.

Medicinal Parts: Whole plant.

Chemical constituents: Swertiamarin, gentianine, apigenin, genkwanin, Isovitexin, saponarin, heptacosane, nonacosane, myristic acid, oleic acid, gentiocrucine, alkaloids.

Medicinal Properties: Acrid, anthelmintic, anti – inflammatory, antiperiodic, bitter, carminative, depurative, digestive, stomachic, thermogenic.

Medicinal Uses: Constipating, dyspepsia, dropsy, fever leprosy, pruritus, skin diseases, swelling.

177. *Erythrina variegata* L. (Fabaceae)

Syn.: *E. indica* Lam.

Vern.: Paribhadra.

Distribution: Planted as an ornamental. Bhopal, Indore, Seoni.

Botany: Small tree, with 3-foliate leaves, red flowers and black, torulose pods.

Flowering & Fruiting: March-September.

Medicinal Parts: Bark, Leaves.

Chemical constituents: Bark contains A bitter erytherine alkaloid & resins.

Leaves contain Erytherine, alkaloid.

Medicinal Properties: Bark – Acrid, anthelmintic, anti – inflammatory, bitter, carminative, depurative, digestive, diuretic, expectorant, febrifuge, rejuvenating, sedative, stomachic, thermogenic.

Leaf – Anthelmintic, bitter, diuretic, laxative, stomachic.

Medicinal Uses: Bark – Colic, conjunctivitis, cough, dyspepsia, fevers, inflammation, insomnia, leprosy, otalgia, skin diseases.

Leaf – Dysmenorrhoea, otalgia.

Ayurvedic preperations: *Paribhadra-avleha.*

178. *Eulophia nuda* Lindl. (Orchidaceae)

Distribution: In Sal forests, alongwith *Elephantopus scaber.* Balaghat.

Botany: Terrestrial orchid; with tuberous roots & Whitish-Purple flowers in spadix inflorescence.

Flowering & Fruiting: April-September.

Medicinal parts: Tubers.

Chemical constituents: *Tuber contains* – Eulophiol, n-hexacosyl alcohol, Nudol, Lupeol.

Medicinal properties: *Tuber* – Antitumour, Bronchodilator, Vermifuge.

Medicinal uses: *Tuber* – Intestinal worms, Tumours, Bronchitis, Diseases of the Blood.

179. *Euphorbia antiquorum* L. (Euphorbiaceae)

Vern.: Tridhara.

Distribution: Rare. In Rock crevices. Hoshangabad.

Botany: Much branched shrubs or a small trees; branches curving upwards, stout, fleshy, jointed, 3-angled; leaves subsessile, obovate-oblong or spathulate, fleshy, long, Deciduous; involucres 3-nate, forming small pedunculate cymes, central flowers sessile, lateral stalked; cocci compressed, glabrous.

Flowering & Fruiting: November-January.

Medicinal Parts: Root, Juice.

Chemical constituents: Plant contains Euphol, Taraxerol, Epifriedelanol, friedelan.

Latex cntains β – Amyrin, antiquorin, cycloartenol, œ – euphorbol, euphadienol.

Medicinal Properties: Root – Acrid, anodyne, bitter, digestive, emetic, purgative, stomachic, thermogenic.

Juice – Acrid, anti-Inflammatory, purgative.

Medicinal Uses: Root – Colic, constipation, dyspepsia, ulcers, wounds.

Juice – Cough, deafness, dropsy, gout, rheumatism.

180. *Euphorbia hirta* L. (Euphorbiaceae)

Syn.: *E. pilulifera* L.

Vern.: Dudhi.

Distribution: Common along roadsides and waste moist places. Balaghat, Chhatarpur, Damoh, Hosahangabad, Indore, Jabalpur, Satna, Tikamgarh.

Botany: Erect or procumbent herbs, with ovate-lanceolate or rhomboid-oblong leaves. Cyathia greenish, crowded in pedunculate cymes; glands stalked, pink. Capsules globosely 3-gonous.

Flowering & Fruiting: Throghout the year.

Medicinal Parts: Whole plant.

Chemical constituents: Plant contains Quercetin, Triacentane, ellagic acid, Phenolics, Euphosterol, Phytosterol, Phytosterolin, Gallic, mellistic, palmitic, oleic, linoleic acid, zanthorhamnin wax, chlorophyll, resin, tannins, sugar, mucilage, carbohydrates, albuminoids, catchin, oxalate.

Medicinal Properties: Astringent, diuretic, expectorant.

Medicinal Uses: Asthma, colic, cough, dysentery, gonorrhoea, intestinal worms in children, leucoderma.

181. *Euphorbia neriifolia* L. (Euphorbiaceae)

Vern.: Pattonkisend.

Distribution: Usually in drier or rocky regions, wastelands and often planted as hedge around the villages and cultivated fields. Damoh, E. Nimar, Hosahangabad, Panna, Seoni, Sidhi, Tikamgarh.

Botany: Erect deciduous shrubs or small trees; stem 5-angled. Leaves obovate-oblong, crowded at the end of stem, subfleshy. Involucre 3-nate; glands without petaloid limb. Capsules obovoid-trigonous, deep red or pale brown.

Flowering & Fruiting: November-December.

Medicinal Parts: Roots, stem, leaves.

Chemical constituents: Plant contains Euphol, taraxerol, friedelan, resin, gum.

Bark contains n – Hexacosonal, Euphorbol, Hexacos

Anoate, Pelargonine, diglusoside, Euphorbin.

Root & bark contains Methylene Cycloartenol, Tulipanin 3-5 diglucoside.

Medicinal Properties: Plant – Alexipharmic, bitter, carminative, laxative, pungent.

Milk – Appetiser, laxative, pungent.

Leaf – Appetiser, carminative.

Medicinal Uses: Plant – Abdominal trouble, anaemia, bronchitis, fever, inflammation, leucoderma, piles, tumours, ulcer.

Milk – Abdominal pain, leucoderma, tumor.

Leaf – Abdominal trouble, inflammation, swelling, tumour.

Ayurvedic preperations: *Snhuadi varti, Snhusdi taila, Vajrakshar.*

182. *Euphorbia nivulia* Buch-Ham. (Euphorbiaceae)

Vern.: *Katathohar.*

Distribution: Common on slopes & foot hills. Chhatarpur, Damoh, Mandla, Rewa, Satna, Seoni.

Botany: Tall shrubs or small trees with whorled cylindrical fleshy stems and branches, covered with pairs of straight spines, leaves 15 cm. or more long.

Flowering & Fruiting: March-May.

Medicinal parts: Root, Bark, Latex, Stem.

Chemical constituents: *Stem* – Taraxerol, Sitosterol.

Root bark – Euphol, Euphorbol, Hexacozoate.

Latex – Cyclolaudenol, Cyclonivuliaol, Cycloartenol, Cycloeucalenol.

Medicinal properties: *Stem* – Antirabies.

Root bark – Antidropsial.

Latex – Abortifacient, Bitter, Diuretic, Digestive, Expectorant, Purgative, Rubefacient, Stomachic.

Medicinal uses: *Latex* – Asthma, Bitter, Diuretic, Digestive, Earache, Flatulence, Intermittent fever, Jaundice, Leprosy, Rheumatism, Scabies, Warts, Ulcers.

183. *Euphorbia thymifolia* L. (Euphorbiaceae)

Distribution: Common in moist places and along the roads and ridges. Balaghat, Damoh, E. Nimar, Indore, Mandla, Panna, Rewa, Sagar.

Botany: An annual herb. Leaves very small and opposite. Involucers axillary, solitary or 2-3 in an axil. Capsules erect, obtusely keeled; styles short, 2-fid.

Flowering & Fruiting: Throughout the year.

Medicinal Parts: Whole plant.

Chemical Constituents: Plant contains Epitaraxerol, N-hexacosanol, Euphorbol, Esters, N-alkanes, Sterols, Corilagin, Geraniin, Bixanin, Casuarin, Desgalloylstachyurin.

Medicinal Properties: Acrid, Alexipharmic, Anthelmintic, Antibacterial, Antidiarrhoeal, Aphrodisiac, Astringent,, Bitter, Demulcent, Depurative, Diuretic,

Expectorant, Laxative, Purgative, Vulnerary.

Medicinal Uses: Amenorrhoea, Asthma, Bronchitis, Bleeding, Piles, Cardiac debility, Chronic Cough, Dandruff, Dysentery, Dysmenorrhoea, Flatulence, Gonorrhoea, Helminthiasis, Leprosy, Skin diseases, Strangury, Wounds.

184. *Euphorbia tirucalli* L. (Euphorbiaceae)

Vern.: Thor.

Distribution: Rare. In drier places of hilly or rocky regions. Chhatarpur, Damoh, Indore, Tikamgarh.

Botany: Shrubs or small trees, with smooth, spreading, terete, rush-like, almost leafless branches.

Flowering & Fruiting: August– September.

Medicinal Parts: Root, stem, latex, bark.

Chemical constituents: Stem & latex contain β – Sterol, Taraxerol.

Bark & root contain Euphorbol, Hexacosanoate, Taraxerone, Ingenol, Triacetate, Euphorginol.

Stem contain Ellagic acid, Kaempferol, compesterol, Stigma – Sterol.

Latex contain Amyrin, Euphol, Euphorbinol, Tirucallol, Palmitic acid.

Medicinal Properties: Milk – Alexiteric, carminative, emetic, purgative, rubefacient.

Medicinal Uses: Plant – Biliousness, leprosy, leucorrhoea.

Milk – Abdominal trouble, tumour.

185. *Evolvulus alsinoides* L. (Convolvulaceae)

Syn.: *E. hirsutus* Lamk.

Vern.: Nil Shankhpuspi.

Distribution: Common in wastelands and dry mixed deciduous forests, mostly in rockey and dry places. Balaghat, Chhatarpur, Damoh, Indore, Rewa, Satna, Seoni, Shivpuri, Sidhi, W. Nimar.

Botany: Prostrate or ascending herbs, with lanceolate-oblong leaves. Flowers pale-blue, on 1 to few flowered filiform peduncles. Calyx-lobes acute. Capsules 4-valved, 4-seeded.

Flowering & Fruiting: July – March.

Medicinal Parts: Whole plant.

Chemical constituents: Plant contains Yellow fatty elements, Minerals, Organic acid & Saline.

Medicinal Properties: Plant – Alexiteric, alterative, anthelmintic, bitter, pungent, tonic.

Medicinal Uses: Plant – Biliousness, bronchitis, epilepsy, leucoderma, teething in infants.

Ayurvedic preperations: *Sankhapuspi panak, Sarasvata churna, Amritadi rasayan, Medhya kasaya, Brahmi rasayana, Medhya rasayana, Indra rasayana.*

186. *Exacum bicolor* Roxb. (Gentianaceae)

Vern.: **Baracharayata.**

Distribution: **In moist places.** Hosahangabad.

Botany: An erect annual herb.

Flowering & Fruiting: September-January.

Medicinal parts: Whole plant.

Medicinal properties: Tonic, Stomachic.

Medicinal uses: This drug is used as a substitute for Gentian.

187. *Ficus arnottiana* (Miq.) Miq. (Moraceae)

Vern.: Paras-pipal.

Distribution: Common in mixed forests amidst rocky boulders & along the drains. Mandla, Panna, Satna, Seoni, Shahdol, Sidhi.

Botany: A glabrous shrub or tree; Leaves broadly ovate, base cordate, Margin entire; Stipules ovate-lanceolate; Receptacles globose in pairs or clusters from tubercles in the axils of fallen leaves. Male flowers few, Gall & Fertile flowers indistinguishable.

Flowering & Fruiting: February-April.

Medicinal Parts: Bark.

Medicinal Properties: Bark – Aphrodisiac, Astringent, Constipating, Demulcent, Depurative, Emollient, Refrigerant, Sour, Sweet.

Medicinal Uses: Bark – Burning sensation, Diabetes, Diarrhoea. Inflammation, Leprosy, Pruritus, Scabies, Skin diseases, Wounds.

188. *Ficus benghalensis* L. (Moraceae)

Vern.: Bargad.

Distribution: Common throughout the State on roadsides, in rural areas and secondary forests. Chhatarpur, Damoh, Hosahangabad, Indore, Panna, Rewa, Sagar, Satna, Seoni, Sidhi, Tikamgarh, W. Nimar.

Botany: Large, evergreen trees, with prop roots and ovate-oblong leaves. Receptacles geminate, axillary, globose, red when ripe.

Flowering & Fruiting: August – July.

Medicinal Parts: Whole plant, buds, roots, latex.

Chemical constituents: Plant contains Glucoside; Bengalenoside & Talbutamide; Leucoanthocyanidins, Phytosterolin, β – sitosterol, freidelin, quercetin, Galactoside, ticlic acid, tannin.

Medicinal Properties: Acrid, astringent, sweet.

Medicinal Uses: Biliousness, erysipelas, fever, inflammation, ulcers, vaginal complaint, vomiting.

Ayurvedic preperations: *Nyagrodhadi churna, Nyagrodhadi ghrita, Nyagrodhadi kwath, Majun bargad, Vatajatadi taila.*

189. *Ficus carica* L. (Moraceae)

Vern.: Anjir.

Distribution: On roadsides & often cultivated in gardens for edible figs. Hosahangabad, Indore.

Botany: A small or middle sized deciduous tree.

Flowering & Fruiting: January-October.

Medicinal parts: Fruits, Roots, Bark, Latex.

Chemical constituents: Proteose, Amino acid, Tyrosin, Enzyme Cravin, Lipase, Protease, Carotin.

Leaves yield bitter substance ficusin & bergaptene.

Latex-Ficin.

Medicinal properties: *Fruit* – Nutritive, Emollient, Demulcent, Aperient, Laxative, Nutritious.

Latex: Anthelminitic.

Medicinal uses: *Fruits* – Kidney & Bladder Stones, Obstruction of the Liver & Spleen, Piles, Gout.

Latex: Mouth ulcers.

190. *Ficus glomerata Roxb.* (Moraceae)

Syn.: *F. racemosa* L.

Vern.: Gular.

Distribution: In wastelands and mixed forests, on the bank of drains and rivers. Balaghat, Chhatarpur, Damoh, Dhar, Hosahangabad, Jabalpur, Jhabua, Mandla, Panna, Rewa, Sagar, Satna, Seoni, Shivpuri, Sidhi, W. Nimar.

Botany: A large tree; leaves alternate, ovate, elliptic-lanceolate, or ovate-oblong, glabrous. Receptacles 2-4 cm across, clustered on short leafless branchlets, globose or pyriform, hairy, supported by 3 bracts at the base, green, turning reddish-purple when ripe.

Flowering & Fruiting: February – June.

Medicinal Parts: Whole plant, roots, leaves, bark, fruits, latex.

Chemical constituents: Bark contains Cerylbehenate, Lupeol, Alphaamyrin, Glaunol, β – sitosterol, keton, Hentriacontane, tiglic acid, glucose, Taraxeroll & Tannin.

Medicinal Properties: Plant – Acrid, antidysenteric, cooling, sweet.

Medicinal Uses: Root – Hydrophobia.

Fruits – Biliousness, blood diseases, burning sensation, fatigue, leprosy, menorrhagia, nose bleeding, urinary discharges.

Ayurvedic preperations: *Udumbar sara, Udumbar avaleha, Udumbaramrta.*

191. *Ficus microcarpa* L. f. (Moraceae)

Syn.: *F. retusa* Auct. non. L.

Distribution: Common in mixed forests, on hill slopes and in rock crevices. Chhatarpur, Mandla, Panna, Sagar, Seoni.

Botany: A large glabrous evergreen tree with few aerial roots; leaves short petioled, stipules lanceolate; fruit receptacles sessile and globose in axillary pairs, yellowish when ripe.

Medicinal Parts: Root, Bark, Leaves.

Medicinal Properties: Bark & Leaves – Acrid, astringent, refrigerant, stomachic.

Medicinal Uses: Bark & Leaves – Burning sensation, diabetes, diarrhoea, dropsy, dysentery, erysipelas, leucorrhoea, ulcers, wounds.

192. *Ficus religiosa* L. (Moraceae)

Vern.: Pipal.

Distribution: Common throughout the State on road sides, bank of drains and rivers, in fields, villages and sometimes in forests. Balaghat, Bhopal, Chhatarpur, Damoh, Indore, Jabalpur, Panna, Shivpuri, Sidhi, W. Nimar.

Botany: Large, deciduous trees, with ovate to ovate-suborbicular leaves caudate-acuminate at apex and rounded at the base. Receptacles in axillary pairs, reddish-purple when ripe.

Flowering & Fruiting: April –September.

Medicinal Parts: Bark, shoot, leaves, fruit.

Chemical constituents: Bark contains Tannin.

Medicinal Properties: Plant – Acrid, bitter, sweet.

Fruits – Alexipharmic, cooling.

Medicinal Uses: Burning sensation, biliousness, diseases of the blood, uterus, vagina; leucorrhoea, ulcers.

Fruits – Biliousness, burning sensation, diseases of the blood & heart.

Root – gout

Root bark – Leucorrhoea, stomatitis, ulcers.

Seed – Urinary discharges.

Ayurvedic preperations: *Hartal bhasm, Sharbat peepar.*

193. *Flacourtia indica* (Burm. f.) Merr. (Flacourtiaceae)

Syn.: *F. ramantchi* L. Herit.; *F. ramontchi* L. Herit.; *F. intermis* auct. (non Roxb.); *F. latifolia* Cooke; *F. occidentalis* Blatter; *F. cataphracta.; F. sapida* Roxb.; *F. sepiaria* Roxb.

Vern.: Bilanga.

Distribution: *In scrub forests. In almost all the Districts.*

Botany: Deciduous thorny trees, with ovate-elliptic or suborbicular leaves and greenish-yellow flowers.

Flowering & Fruiting: March-May.

Medicinal Parts: Root, Leaves, Fruits.

Taste: Sweet.

Chemical constituents: Bark contains Tannin & flacourtin (A Phenolic glucoside ester).

Medicinal Properties: Roots – Alexipharmic, depurative, diuretic, refrigerant, sweet.

Fruits – Appetizer, digestive, diuretic, sweet.

Medicinal Uses: Roots – Erysipelas, poisonous bites, pruritus, skin diseases, strangury.

Leaves – Pruritus, scabies.

Fruits – Jaundice, strangury.

194 *Flacourtia jangomas* Rausch. (Flacourtiaceae)

Syn.: *F. cataphracta* Roxb.

Vern.: *Panialah.*

Distribution: In Sal forests. In almost all the districts.

Botany: Large deciduous shrub or trees, with simple, ovate, glabrous leaves and Racemose Inflorescence. *Flowering & Fruiting:* June-December.

Medicinal Parts: Bark, Leaves, Fruits.

Chemical Constituents: Ripe fruits – Reducing Sugars, Ash, Fat, Potassium, Tartaric acid, Proteins, Phenolics, Nitrogen, Iron, Phosphorus, Ascrobic acid and Unidentified chemicals. Gas chromatographic analysis of Fatty acids in Fats revealed the presence of palmitic, hexadecadienoic, Stearic, Oleic, Linoleic, □ -linoleic and a few minor unidentifed acids. Amino acids from extract of their-dried ripe fruits showed the presence of Proline Hydroxyproline, Methtonine, Alanine, Glycine, and Valine. Paper Chromatography studies on simple reducing Sugars and their alditol acetates indicated the presence of Arabinose, Glucose, Fructose and Galactose.

Medicinal Properties: Bark, Leaves – Astringent, Acrid, Sour, Refrigerant, Stomachic, Diaphoretic.

Fruits – Sweet, Sour, Astringent, Acrid, Refrigerant, Digestive, Stomachic, Alexipharmic, Anti-Inflammatory, Depurative, Liver tonic.

Medicinal Uses: Bark, Leaves – Debility, Diarrhoea, Haemorrhoids, Stomatitis.

Fruits – Colic, Diabetes, Dyspepsia, Flatulence, Inflammation, Jaundice, Nausea, Rheumatism, Skin diseases.

195. *Foeniculum vulgare* Mill. (Apiaceae)

Vern.: Saunf.

Syn.: *F. capillaceus* Gilb.; *Anethum foeniculum* L.; *A. officinale* All.

Distribution: Often cultivated for fruits which are used as spices.

Botany: A large annual herb, pinnately compound leaves; Flowers yellow in compound umbels; cremocarp of 2 carples or mericarp, each one in 5-ridged furrowed.

Flowering & Fruiting: December-March.

Medicinal Parts: Fruits.

Chemical constituents: Constituents of plant are Fatty oil, protein, anethose, fenchone, anisalehyde, feniculin, nelumboside; Flavons, abscisic acid. Essential oil from fruit anethole, xanthotoxin, bergapten, psoralen, scoparone, vanillin, α-pinene, camphene, limonene, α-phellandrene, fenchone, methylchavicol.

Medicinal Properties: Fruits – Acrid, alexipharmic, anthelmintic, antiemetic, aphrodisiac, bitter, carminative, digestive, diuretic, emollient, expectorant, febrifuge, haematinic, ophthalmic, refrigerant, stomachic, sudorific, sweet.

Medicinal Uses: Fruits – Anaemia, bruning sensation, bronchitis, cardiac diseases, colic, cough, dysentery, dyspepsia, dysuria, fever, flatulence, inflammation, skin diseases, vomiting.

196. *Fumaria indica* (Hassk.) Pugsley (Fumariaceae)

Syn.: *F. vaillantii* Loisel.; *F. parviflora* Lam.

Vern.: Pitpapra.

Distribution: A weed in cultivated fields. Chhatarpur, Damoh, Satna, Shahdol.

Botany: Annual herbs; leaves 2-3 pinnatisect, segments linear-oblanceolate; flowers purplish-red, spurred, in terminal or leaf opposed, bracteate racemes; sepals inciso-dentate; nuts subglobose, rugose when dry.

Flowering & Fruiting: January – April.

Medicinal Parts: Whole plant.

Chemical constituents: Plant contains Fumaric acid & fumarin alkaloid.

Medicinal Properties: Aperient, astringent, bitter, blood purifier, diaphoretic, diuretic, tonic.

Medicinal Uses: Blood diseases, fever, leprosy, rhinitis, tuberculousis, vertigo.

Ayurvedic preperations: *Parpati kwath, Parpatadhyarista, Panchbhadra kwath, Shadgapaniya, Satayadi kwath, Kirattiktadi kwath, Sudarshan churna, Amritarista, Bhunimbadi vati.*

197. *Gardenia gummifera* L. f. (Rubiaceae)

Vern.: Chhote kurra.

Syn.: *G. lucida* Roxb.

Distribution: Balaghat, Seoni.

Botany: A small compact tree; leaves simple smooth & bright; flowers white during early period but yellow later on; 1-3 in cluster, gum is secreted from tender branches and buds during winter.

Flowering & Fruiting: March-August.

Medicinal Parts: Gum.

Chemical constituents: Oleanonic aldehyde, sitosterol, D-mannitol, erythrodiol, 5-flavons, gardenin, de-me-tangeretin nevadensin.

Fruits contain Nonacosane, β-sitosterol, d-mannitol.

Medicinal properties: Resin – Acrid, anodyne, anthelmintic, antispasmodic, appetiser, bitter, carminative, depurative. diaphoretic, digestive, expectorant, thermogenic, vulnerary.

Medicinal Uses: Resin – Bronchitis, cardiac debility, colic, constipation, cough, dyspepsia, fever, flatulence, hiccough, intestinal worms, leprosy, skin diseases, wounds

198. *Garuga pinnata* Roxb. (Burseraceae)

Vern.: *Kekad.*

Distribution: Bhopal, Mandla, Seoni.

Botany: Large deciduous trees, with creamy White flowers and fleshy, glabrous, subpyriform drupes.

Flowering & Fruiting: February-June.

Medicinal Parts: Root, Stem, Leaves, Fruits, Galls.

Medicinal Properties: Root – Pulmonary affections.

Fruits – Carminative, Cooling, Digestive, Sweet, Vermifuge.

Medicinal Uses: Leaf – Asthma.

Fruits – Gastropathy, Roundworm.

Galls – Foul ulcers, Obesity, Splenomegaly.

199. *Glinus oppositifolius* DC. (Molluginaceae)

Syn.: *Molluga oppositifolia* L.; *M. spergula* L.

Distribution: Balaghat, Damoh, Mandla, Sidhi.

Botany: Prostrate or diffuse herbs with small axillary White flowers & 3 valved capsule fruits.

Flowering & Fruiting: Throughout the year.

Medicinal parts: Whole plant

Chemical constituents: Vitex, Spergulagenic acid, Oleanolic acid, Methyl Spergulagenate, Spergulagenin A.

Medicinal properties: Antiseptic, Aperient, Anthelmintic, Diuretic, Stomachache.

Medicinal uses: Burning sensation, Bronchitis, Cough, Diabetes, Earache, Piles, Skin diseases.

200. *Gloriosa superba* L. (Liliaceae)

Syn.: *Methonia superba* Lamk.

Vern.: Kalihari.

Distribution: Abundant on bushes in open forests. Balaghat, Damoh, Dhar, Hosahangabad, Indore, Mandla, Satna, Seoni, Sidhi, Tikamgarh.

Botany: Climbing herbs, with alternate, opposite or whorled leaves ending in a tendril. Flowers yellow, tinged with red in upper half; tepals undulating on the margins, Capsules oblong.

Flowering & Fruiting: June-November.

Medicinal Parts: Rhizome, root.

Chemical constituents: Rhizome tubers contain Alkaloid colchicine; superbine, glorisine; sitosterol, glucosides, benzoic acid & resins.

Medicinal Properties: Tuber – Abortifacient, acrid, alexiteric anthelmintic, bitter, heating, laxative, Pungent.

Medicinal Uses: Tuber – Abdominal pain, infalammation, itching, leprosy, piles, ulcer.

201. *Gmelina arborea* Roxb. (Verbenaceae)

Syn.: *G. rheedii* Hook.

Vern.: Kamhar.

Distribution: In sal forests. Chhatarpur, Damoh, Hosahangabad, Indore, Mandla, Seoni, Shahdol, Shivpuri.

Botany: Unarmed deciduous trees, with broadly ovate-acuminate leaves. Flowers yellow or brownish-yellow. Drupes obovoid, yellowish orange when ripe.

Flowering & Fruiting: March-June.

Medicinal Parts: Whole plant, roots, leaves, fruits.

Chemical constituents: Plant contains Lignan gmelinol & Arboreol.

Leaves contains Luteolin, Apigenin, Quercetin, β – Sitosterol, Isoarboreal, Oleic acid, Gmelanon, Gummadiol, Palmitic acid, oleic acid,Linoleic acid, Glucose, Xylose, Mannose, Arabinose, Galactose, Rhamnose, Stigmasterol, Sitgmastanol, Compestrol, Butulinol.

Medicinal Properties: Root – Acrid, anthelmintic, appetiser, bitter, indigestion, laxative stomachic, sweet.

Flower – Astringent, bitter, cooling, sweet.

Fruits – Alterative, astringent, bitter, cooling, sweet.

Medicinal Uses: Root – Abdominal pain, burning sensation, fever, hallucinations, piles, urinary discharges.

Flower – Blood diseases, leprosy.

Fruits – Anaemia, leprosy, strangury, ulcers, vaginal discharges.

Ayurvedic preperations: *Dashmularista, Kutajarisht, Sriparnyadi kwath, sriparni taila, Brihat panchmuladi kwath.*

202. *Gossypium arboreum* L. (Malvaceae)

Syn.: *G. indicum* Tod.; *G. neglectum* Tod.; *G. intermedium* Tod.; *G. cernuum* Tod.; *G. nanking* Meyen.

Distribution: Cultivated for cotton. Shivpuri, Sidhi.

Botany: A large perennial shrub or small tree; leaves simple, alternate, ovate-cordate, palmately 3-7 lobed; flowers purple with darker coloured claws; fruits capsules.

Flowering & Fruiting: January-April.

Medicinal Parts: Leaves.

Chemical constituents: Alcoholic extract of aerial parts is CNS depressant.

Medicinal Properties: Leaves – Anti-inflammatory, astringent, bitter, demulcent, depurative, diuretic, febrifuge, thermogenic, vulnerary.

Medicinal Uses: Leaves – Amenorrhoea, dysmenorrhoea, erysipelas, fever, inflammations, leprosy, otalgia, pruritus, skin diseases, strangury, ulcers, wounds.

203. *Grewia subinaequalis* DC. (Tiliaceae)

Syn.: *G. hainesiana* Hole.

Vern.: Phalsa.

Distribution: Occasional in Sal forest. Balaghat, Hosahangabad, Satna.

Botany: Shrubs or small trees, with ovate to orbicular leaves and yellow flowers in extra-axillary cymes. Petals half the length of basal glands. Nuts 1-celled.

Flowering & Fruiting: November – May.

Medicinal Parts: Root bark, Fruits, Leaves.

Chemical constituents: Plant contains Lupeol, Lupenon, Betulin, Friedelin, Taraxasterol, Beta-Amyrine, Grewionol, Kampferol, Amino acids, Citric acid, Vitamin-C.

Medicinal Properties: Fruits – Aphrodisiac, appetiser, burning sensation, cooling, digestive, sweet, tonic.

Medicinal Uses: Fruits – Blood disorders, fever, heart disorders, inflammation, oedema, stomachahe.

204. *Grewia tenax* fiori. (Tiliaceae)

Syn.: *G. populifolia* Vahl.

Distribution: In mixed forests.

Botany: Shrub, with Yellow flowers & drupe fruits.

Flowering & Fruiting: May-September.

Medicinal parts: Root bark & Fruits.

Chemical constituents: *Stem bark & Leaf* – Triacontan-1-ol, □ -Amyrin, β-Sitosterol, Lupenone, Erythrodiol, Betulin, Tetratriacont-21-ol-12-one.

Medicinal properties:

Medicinal uses: Dysentery, Bone fracture, Pain, Wound healing.

205. *Grewia tiliifolia Vahl.* (Tiliaceae)

Vern.: *Dhaman.*

Distribution: In mixed & sal forests. Balaghat, Hosahangabad, Rewa,, Seoni, Shahdol, Shivpuri.

Botany: Small trees, with ovate to orbicular leaves and yellow flowers in axillary cymes. Drupes globose.

Flowering & Fruiting: May-January.

Medicinal Parts: Bark, Fruits.

Chemical Constituents: Leaves contain – Tannin.

Medicinal Properties: Bark – Acrid, Antipruritic, Aphrodisiac, Astringent, Constipating, Emetic, Expectorant, Refrigerant, Styptic, Sweet, Tonic, Vulnerary.

Medicinal Uses: Bark – Burning sensation, Cough, Diarrhoea, General debility, Haemorrhages, Pruritus, Seminal weakness, Skin diseases, Ulcers, Wounds.

206. *Gymnema sylvestre* (Retz.) R. Br. ex Sch. (Asclepiadaceae)

Syn.: *Asclepias geminata* Roxb.; *Periploca sylvestris* Retz.

Vern.: Gurmar.

Distribution: In sal and mixed deciduous forests. Betul, Chhatarpur, Damoh, Hosahangabad, Jabalpur, Narsimhapur, Panna, Raisen, Rewa, Satna, Shahdol.

Botany: Much branched, twining shrubs. Leaves ovate-elliptic or oblong. Cymes greenish–yellow. Follicles solitary, straight, lanceolate, tapering.

Flowering & Fruiting: October-February.

Medicinal Parts: Whole plant.

Chemical constituents: Leaves contains Gymnemic acid, quercitol, anthraquinon, saponin, calcium oxalate.

Medicinal Properties: Plant – Acrid, alterative, alexiteric anthelmintic, bitter, cooling, tonic.

Fruits – anthelmintic, alexiteric, bitter, pungent, stomachic.

Medicinal Uses: Plant – Asthma, biliousness, bronchitis, burning sensation, eye complaint, heart diseases, inflammation, leucoderma, ulcer.

Fruits – Bronchitis, heart diseases, ulcers, urinary discharges.

207. *Haldina cordifolia* (Roxb.) Ridsd. (. (Rubiaceae)

Syn.: *Adina cordifolia* Roxb.) Hook. f. ex Bran

Vern.: Haldu

Distribution: Balaghat, Betul, Chhatarpur, Damoh, Gwalior, Indore, Mandla, Panna, Satna, Seoni, Shahdol, Sidhi.

Botany: Large tree with spreading branches; leaves elliptic-oblong; flowers sweet scented; fruits fleshy.

Flowering & Fruiting: August-June.

Medicinal Parts: Root, Stem bark.

Chemical constituents: Fruit contains Flavonoids.

Leaves contain Ursolic acid, quercetin 3–O– galactoside.

Heartwood contains Adinin pigment, tannins.

Stem bark contains Alkaloids, tannins pentosan, lignins, starch.

Root bark contains Coumarin glycoside adicardin.

Medicinal Properties: Root – Astringent

Bark – Acrid, antiseptic, aphrodisiac, Bitter, Febrifuge, Pungent, tonic, vulnerary.

Medicinal Uses: Root – Dysentery.

Bark – Diseases of blood, inflammation, skin diseases.

208. *Hardwickia pinnata* Roxb. (Caesalpinniaceae)

Syn.: *Kingiodindron pinnatum* Harm.

Vern.: Anjan.

Distribution: In dry deciduous forests. Bhopal, Gwalior, Hosahangabad, E. Nimar, Indore.

Botany: Evergreen tall trees, with small dull Yellow flowers & single seeded pod fruits.

Flowering & Fruiting: October-April.

Medicinal parts: Resin.

Chemical constituents: *Resin* – Sesquiterpens, Diterpenes, Caryophyllene alcohol, Humulene Oxide-I & II, Caryophyllenol-I & II, β-Sitosterol, Clovanecriol, Kolavenol, Kolavelool, Kolavonic acid, Kolavenic acid, Kolavenolic acid, Hardwickiic acid, Kolavic acid, Cadinene, β-Caryophyllene, α-Caryophyllene.

Medicinal properties: *Resin* – Antigonnorrhoic, Anti-rheumatic, Diuretic.

Medicinal uses: *Resin* – Rheumatism.

209. *Hedychium coronarium Koen.* (Zingiberaceae)

Distribution: In moist places. Seoni.

Botany: Herbaceous, rhizomatous robust plants, with White or Yellow flowers & capsule fruits.

Flowering & Fruiting: August-February.

Medicinal parts: Rhizomes, Flowers.

Chemical constituents: *Rhizomes* – Furanoditerpene as hedychenone.

Essential oil – β-Pinene, β-Pinene, Limonene, β-Phellandrene, β-Caryophyllene.

Medicinal properties: *Rhizomes* – Anti-rheumatic, Anthelmintic, Carminative, Stimulant, Tonic.

Medicinal uses: *Rhizomes* – Intestinal worms, Fever, Rheumatism, Tonic.

210. *Hedyotis herbacea* L. (Rubiaceae)

Syn.: *Oldenlandia herbacea* (L.) Roxb.

Distribution: Hosahangabad, Seoni, Sidhi.

Botany: An erect bushy annual or biennial; leaves simple, linear-lanceolate, opposite; flowers White, small and fruits capsules.

Flowering & Fruiting: October-December.

Medicinal Parts: Whole plant.

Chemical Constituents: Kaempferol-3-O-Rutinoside and ursolic acid were Isolated from the leaves & tiwgs.

Medicinal Properties: Plant – Anthelmintic, Anti-Inflammatory, Bitter, Cooling, Expecotrant, Febrifuge, Stomachic, Sweet.

Medicinal Uses: Plant – Bronchitis, Elephantiasis, Fever, Hydrocele, Ulcers.

211. *Helianthus annus* L. (Asteraceae)

Distribution: Cultivated throughout the State for extraction of edible seed oil.

Botany: A small annual herb of about 4-8 feet high; leaves hairy, serrate; flowers in large heads, yellow; fruits cypsella without pappus.

Medicinal Parts: Roots, leaves, flowers, seeds.

Chemical constituents: Seed contains – Protein, fats, carbohydrates, fibres, oil which contains – Liolenic acid.

Leaves contain – Malonic, lactic, succinic, aconitic, fumaric acid.

Medicinal Properties: Roots – Anodyne.

Leaves – Emetic.

Flowers – Acrid, anthelmintic, antiperiodic, aphrodisiac, bitter, refrigerant.

Seeds – Diuretic, expectorant.

Medicinal Uses: Roots – Toothache.

Leaves – Burning sensation, fever, lumbago, malaria ulcers, wounds.

Flowers – Anaemica, asthama, bronchitis, buring sensation, fevers, helminthiasis, hysteria, inflammation, leprosy, pruritus, skin diseases, ulcers.

Seeds – Catarrh, cough, strangury.

212. *Helicteres isora* L. (Sterculiaceae)

Vern.: Marod phali.

Distribution: Common in deciduous forests and open areas. Balaghat, Damoh, Hosahangabad, Indore, Narsimhapur, Panna, Sagar, Seoni, Tikamgarh.

Botany: Shrubs, with ovate to orbicular leaves and red flowers. Fruits cylindric, woody, beaked, with spirally coiled follicles.

Flowering & Fruiting: August-June.

Medicinal Parts: Root, Leaf, Stem bark, seed, fruit.

Medicinal Properties: Fruit – Astringent, demulcent, expectorant.

Medicinal Uses: Diarrhoea, dysentery, scabies, stomach affections.

Ayurvedic preperations: *Gandharva churna, Sidha praneshwar ras.*

213. *Heliotropium indicum* L. (Boraginaceae)

Vern.: *Hathisunda.*

Distribution: In moist wastelands. Balaghat, Damoh, Mandla.

Botany: Erect herbs, pale-violet flowers and fruit of 1-seeded nutlets. *Flowering & Fruiting:* Throughout the year.

Medicinal Parts: Whole plant.

Chemical Constituents: Aerial parts of the plant contains as Indicine, Echinatine, Supinine, Heleurine, Heliotrine, Lasiocarpine, Supinidine, Retranecine, Lindelofidine, Indicine-N-oxide, Indicinine, Supinene, Rapanone, Lupeol, N-hexacosanol, Sitosterol, Stigmasterol, Campesterol, Cholinasterol.

Medicinal Properties: Plant – Astringent, Bitter, Diuretic, Emollient, Thermogenic, Vulnerary.

Root – Aphrodisiac, Astringent, Bitter, Expectorant, Febrifuge.

Medicinal Uses: Plant – Wounds, Skin affections, Rheumatism, Ulcers.

Root – Cough, Bornchitis, Fever, Ringworm, Rheumatism.

Leaves – Fever, Urticaria, Ulcers, Wounds, Gonorrhoea, Ringworm, Rheumatism, Erysipelas.

214. *Hemidesmus indicus* (L.) R. Br. (Asclepiadaceae)

Syn.: *Asclepias pseudosarsa* var. *latifolia* Roxb.; *Periploca indica* Willd.; *P. emetica* Retz

Vern.: Anantmul.

Distribution: On forest floor and twiner on a number of Plants. Betul, Chhatarpur, Hosahangabad, Panna, Seoni, Sidhi.

Botany: Glabrous, scandent undershrubs, with elliptic-oblong to linear-lanceolate leaves. Flowers dull yellow with purple eye. Follicles tapering to a point. Slightly falcate.

Flowering & Fruiting: August-January.

Medicinal Parts: Root.

Chemical constituents: Root contains Essential oil, saponin, hemidester, hemidesmol, β – sitosterol, lupeol, tetracyclictriserpene, alcohol, resin, vasa acid, tannins.

Medicinal Properties: Root – Alexiteric, antidiarrhoea, antipyretic, aphrodisiac, astringent, bitter, cooling, sweet.

Medicinal Uses: Root – Asthma, diarrhoea, blood diseases, burning sensation, bronchitis, dysentery, fever, itching, leprosy, leucoderma, leucorrhoea, loss of appetite, piles, skin diseases.

Ayurvedic preperations: *Sarivadi kwath, Sarivadi vati, Sarivadaya leh, Sarivadayasav, Pinda taila.*

215. *Hibiscus cannabinus* L. (Malvaceae)

Vern.: *Patsan.*

Distribution: Damoh, Sidhi.

Botany: Annual or perennial, prickly herbs. corolla spreading, yellow with crimson centre.

Flowering & Fruiting: September-November.

Medicinal Parts: Flower, Seed, Leaves.

Chemical Constituents: Seed – Phospholipids, Palmitic acid, Stearic acid, Oleic acid, Linolic acid, β-sitosterol, Limonene, Citral.

Leaves – Aldobiouronic acid, Rutin, Isoquercitrin, Limonene, Phellandrene, β-terpenyl acetate, Lignoceric acid, Cerotic acid, Montanic acid, Melissic acid.

Flower – Anthocyanin, Fructose, Glucose, Mannose, Sucrose, Sterculic acid, Tridecanoic acid.

Medicinal Properties: Seeds – Analgesic, Aphrodisiac, Appetising, Stomachic.

Leaves – Purgative.

Flower – Antibilious, Antipyretic, Antiulcer, Digestive.

Medicinal Uses: Flower – Abdominal pain, Biliousness, Digestion, Dysentery, Fever.

216. *Hibiscus rosa-sinensis* L. (Malvaceae)

Vern.: Gurhal.

Distribution: Commonly grown in the gardens throughout Madhya Pradesh, mostly as a hedge and as an ornamental.

Botany: Evergreen shrubs, with pink, red or rosy flowers.

Flowering & Fruiting: Throughout the year.

Medicinal Parts: Root, stem, leaves, floral buds, flowers.

Chemical constituents: Leaves contain Alkaloids, Glycosides, Reducing sugars, Sterols, Resins, Beta-sitosterol.

Root contain Mucilage, fibres.

Flowers contain Fat, calcium, phosphorous, Thiamine, riboflavin, ascorbic acid, cyanin chloride.

Medicinal Properties: Analgesic, antipyretic, aperient, aphrodisiac, cardiac, contraceptive, demulcent, diuretic, emolient, laxative, refrigerant.

Medicinal Uses: Cough, fever, anti-inflammatery, anti-bacterial, anti-fungal, antispasmodic, anti-viral, gonorrhoea, leucorrhoea.

Ayurvedic preperations: *Japa taila.*

217. Hiptage benghalensis (L.) Kurz (Malpighiaceae)

Syn.: *H. madablota* Gaertn.; *H. parvifolia* W. & A.

Vern.: Madhavilata.

Distribution: Bhopal, Chhatarpur, Rewa, Sidhi.

Botany: A large woody evergreen climbing shrub with silky young parts; leaves simple, opposite, coriaceous, elliptic-oblong, ovate-lanceolate; flowers yellowish-white, fragrant in racemes; fruits samara.

Flowering & Fruiting: January-April.

Medicinal Parts: Bark, leaves, flowers.

Chemical constituents: Root contain – Glucoside hiptagin.

Root bark – Mangiferin.

Medicinal Properties: Bark – Acrid, anti-inflammatory, aromatic, astringent, bitter, expectorant, refrigerant, vulnerary.

Leaves – Acrid, anti-inflammatory, aromatic, astringent, bitter, expectorant, refrigerant, vulnerary.

Flowers – Acrid, anti-inflammatory, aromatic, astringent, bitter, expectorant, refrigerant, vulnerary.

Medicinal Uses: Bark – Asthma, burning sensation, cardiac debility, cough, leprosy, rheumatism. scabies, skin diseases, ulcers, wounds.

Leaves – Asthma, burning sensation, cardiac debility, cough, leprosy, rheumatism. scabies, skin diseases, ulcers, wounds.

Flowers – Asthma, burning sensation, cardiac debility, cough, leprosy, rheumatism. scabies, skin diseases, ulcers, wounds.

218. *Holarrhena pubescens* (Buch.-Ham.) Wall. ex G. Don. (Apocynaceae)

Syn.: *H. antidysenterica* (Roxb. ex f.) Wall.; *Wrightia antidysenterica* J.Grah.; *Echites antidysenterica* Heyne.

Vern.: Kurchi.

Distribution: Common in sal and mixed deciduous forests. Balaghat, Chhatarpur, Damoh, E. Nimar, Gwalior, Hosahangabad, Indore, Jabalpur, Narsimhapur, Panna, Rewa, Satna, Seoni, Shivpuri, Sidhi.

Botany: Small trees, leaves ovate to elliptic. Flowers white, in terminal corymbose cymes, calyx lobes oblong-lanceolate, cilliate, corolla puberulous, follicles cylindric, often dotted with white spots.

Flowering & Fruiting: February – June.

Medicinal Parts: Root, stem bark, seeds..

Chemical constituents: Root bark contains Holacetine, connesine alkaloids.

Seed contains Essential oil, conessine & kurchicine alkaloids.

Stem bark contains Connesine, Holarrhenine, Holarrhimine, kurchine, kurchicine, conamine, & conimine.

Medicinal Properties: Bark – Acrid, anthelmintic, bitter, pungent.

Flowers – Acrid, anthelmintic, appetiser, bitter, cooling.

Seed – Anthelmintic, appetiser, astringent, bitter, cooling, pungent.

Medicinal Uses: Bark – Biliousness, diarrhoea, diseases of spleen, dysentery, fever, leprosy, piles, skin diseases.

Flower – Biliousness, blood diseases, diarrhoea, leucoderma.

Seed – Biliousness, burning sensation, dysentery, erysipelas, hallucinations, leprosy, piles, skin diseases.

Ayurvedic preperations: *Kutaja parpati, Kutajastaka kwath, kutajaghana vati, Kutajarista, Kutajavaleha, Kothari ras, Bhuinimbadi churna, Brihat sudarshan churna, Pathadya, Pradoranilauha, Amritastak kwath, Sudarshan churna, Sarvajwarhar louha, Atsayakadi kwath.*

219. *Holoptelea integrifolia* (Roxb.)Planch. (Ulmaceae)

Vern.: Chilbil.

Distribution: Common on roadsides and around villages. Chhatarpur, Damoh, Indore, Mandla, Sagar, Satna, Seoni, Shivpuri, Sidhi.

Botany: Large trees, with ovate, pinnately veined leaves and greenish-yellow flowers. Fruits brodly oblong to suborbicular, winged, flat, notched at apex.

Flowering & Fruiting: January-April.

Medicinal Parts: Bark.

Medicinal Properties: Bark, Leaves – Acrid, anti-inflammatory, anthelmintic, astringent, bitter, carminative, depurative, laxative, revulsive.

Medicinal Uses: Colic, diabetes, dyspepsia, flatulence, helminthiasis, inflammation, leprosy, rheumatism, skin diseases, vomiting.

220. *Holostemma ada-kodien* Schultes. (Asclepiadaceae)

Syn.: *H. annulare* (Roxb.) K. Schumn.; *H. rheedei* Wall.

Vern.: *Chhirvel.*

Distribution: In sal forest clearings. Hosahangabad.

Botany: Laticiferous twining shrubs, with Purple flowers and foliicles fruits.

Flowering & Fruiting: June-August.

Medicinal Parts: Roots.

Chemical Constituents: Root tubers contain β-amyrin, Lupeol, β-sitosterol, Alanine, Aspartic acid, Glycine, Serine, Thereonine, Valine, Calcium, Phosphate.

Medicinal Properties: Roots – Alterant, Aphrodisiac, Emollient, Expectorant, Ophthalmic, Refrigerant, Stimulant, Sweet, Tonic.

Medicinal Uses: Roots – Burning sensation, Consumption, Cough, Fever, Ophthalmopathy.

221. *Homonoia riparia* Lour. (Euphorbiaceae)

Vern.: *Surra.*

Distribution: Common along the bank of drains & on river beds. Balaghat, Hosahangabad, W. Nimar.

Botany: Deciduous shrubs, with spikes inflorescens and capsule fruits.

Flowering & Fruiting: *January-April.*

Medicinal Parts: Roots, Leaves, Fruits.

Chemical Constituents: Root contains – α-spinasteryl acetate, Fatty acids like myristic acid, Palmitic acid, Stearic acid, Oleic acid.

Leaves contain Quercetain-3-O-β-D-Glucopyranosyl (1→6) – α-L-Rhamnoside, Gallic acid, Taraxerone.

Medicinal Properties: Roots – Depurative,Diuretic, Emetic, Laxative.

Refrigerant, Leaves / Fruits – Antiseptic, Depurative.

Medicinal Uses: Roots – Haemorrhoids, Vesical Calculi, Strangury, Gonorrhoea, Urethrorrhea, Syphilis.

Leaves / Fruits – Ulcers, Wounds, Skin diseases.

222. *Hordeum vulgare* L. (Poaceae)

Syn.: *H. sativum* Pers.

Distribution: Cultivated throughout the State.

Botany: Erect, glabrous annuals, spikes emerging from the leaf-sheaths; spikelets 2-ranked.

Flowering & Fruiting: January-March.

Medicinal Parts: Grains.

Chemical constituents: Fat, starch, protein. Alkaloid hordenine. Luteolin glucoside, orientoside, orientin. Hordenine & its methyl ether.

Medicinal Properties: Grains – Acrid, aphrodisiac, astringent, digestive. diuretic, emollient, refrigerant, sweet.

Medicinal Uses: Grains – Anaemia, asthma, bronchiotitis, burns cough, dyspepsia, fever, stranguary, ulcers.

223. *Hybanthus enneaspermus* F. Muell. (Violaceae)

Syn.: *Viola enneasperma* L.; *V. suffruticosa* L.; *Lonidium suffruticosum* Ging.

Vern.: *Ratanpurus.*

Distribution: Among grasses along roadsides, forest fringes, bank of streams, amongst forest undergrowth, & in cultivated fields.

Botany: Perennial diffuse herbs; flowers Violet; Blue or Orange Red.

Flowering & Fruiting: August-February.

Medicinal parts: Whole plant.

Chemical constituents: Dipeptide alkaloids as Aurantimide acetate, Triterpenes as Isoarborinol & β-Sitosterol.

Medicinal properties: *Plant* – Tonic, Diuretic.

Leaves, Tender stalks – Demulcent.

Root – Antigonorhoic, Diuretic.

Medicinal uses: *Root* – Gonorrhoea, Urinary problems.

Fruits – Antidote for Scorpion sting.

224. *Hygrophila auriculata* (Sch.) Heine (Acanthaceae)

Syn.: *Astercantha longifolia* Nees.

Vern.: Talmakhana.

Distribution: Common along ponds and ditches. Balaghat, Bhopal, Chhatarpur, Dhar, E. Nimar, Hosahangabad, Indore, Jabalpur, Mandla, Narsimhapur, Panna, Rewa, Sagar, Satna, Seoni, Shivpuri, Sidhi, Tikamgarh, W. Nimar.

Botany: Hispid annuals, with thickened nodes. Leaves oblong or lanceolate, in verticils of 6 at a node. Axillary spines yellow. Flowers purple-blue, in axillary whorls. Capsules linear – oblong.

Flowering & Fruiting: September-June.

Medicinal Parts: Roots, leaves, seeds.

Chemical constituents: Seed contains Yellow coloured oil; diastase, lipase, protease, salts of potassium, mucilage.

Medicinal Properties: Latex – Aphrodisiac, bitter, sour, sweet.

Seed – Acrid, aphrodisiac, bitter, constipating, cooling, tonic..

Medicinal Uses: Latex – Abdominal trouble, anaemia, biliousness, constipation, diarrohoea, diseases of the eye, dysentery, inflammation, urinary calculi, urinary discharges.

Seed – Biliousness, blood diseases.

Ayurvedic preperations: *Paushtik churna, Kokilaksa ksara.*

225. *Hyptis suaveolens Poit.* (Lamiaceae)

Vern.: *Dona.*

Distribution: Common as weed in Wastelands, along bank of drains, roadsides, in forest clearings, on hill slopes & among boulders. Chhatarpur, Damoh, Hosahangabad, Mandla, Rewa, Sidhi.

Botany: Herbs; flowers small, Blue, fruits Black nutlets.

Flowering & Fruiting: Throughout the year.

Medicinal parts: Whole plant.

Chemical constituents: Lupeol, Friedelin, Threonine, Valine, Arginine, etc. Amino acids, Essential oil, Campesterol, α-Pinene, Myrcene, β-Caryophyllene.

Medicinal properties: Antiseptic, Carminative, Catarrhal, Stimulant, Sudorific, Galactogogue, Anti-cephalagic.

Leaves – Antispasmodic, Anti-rheumatic, Anti-soporific.

Root – Appetizer, Stomachic.

Medicinal uses: Cough, Corryza, Skin diseases, Epistaxsis.

Leaf juice: Colic, Stomachache

226. *Impatiens balsamina* L. (Balsaminaceae)

Distribution: Balaghat, Damoh, Hosahangabad, Indore, Mandla, Panna, Rewa, Satna.

Botany: Annual herbs, with White or Pink flowers & capsule fruits.

Flowering & Fruiting: July-October.

Medicinal parts: Seeds

Chemical constituents: *Seed contains* – β-Sitosterol, β-amyrin, α-Spinasterol, Glucoside, Linoleic acid, Palmitic acid, Oleic acid, Arachidic acid.

Medicinal properties: The plant is Anti-rheumatic, Cathartic, Emetic.

Medicinal uses: The plant is Amenorrhoea, Urinary disorders, Rheumatism, Vomiting.

227. *Indigofera tinctoria* L. (Fabaceae)

Vern.: Neel.

Distribution: On hill slopes and moist wasteland. Chhindwara, Damoh, Hosahangabad, Indore, Panna, Raisen, Ratlam, Sagar, Satna, Seoni, Shivpuri, Sidhi, Tikamgarh, W. Nimar.

Botany: Grey-pubescent undershrubs, with 7-13, obovate or oblanceolate leaflets. Racemes shorter than leaves. Pods deflexed, 8 to 10 seeded.

Flowering & Fruiting: July-May.

Medicinal Parts: Whole plant.

Chemical constituents: Inorganic salts of Nitrogen, phosphoric acid, lime, potash, apigenin, kaempferol, luteolin, quercetin.

Seeds contain Galactomannan, galactose & mannose.

Medicinal Properties: Root, Leaves & Stem – Anthelmintic, bitter, diuretic, expectorant, laxative.

Medicinal Uses: Root, Stem & Leaves – Asthma, bronchitis, skin diseases, ulcers.]

228. *Ipomoea batatas* (Linn.) Lam. (Convolvulaceae)

Vern.: Shakarkand.

Syn.: *Convolvulus batalas* L.

Distribution: Usually cultivated. Chhatarpur, Damoh, Satna.

Botany: A prostrate vine with tuberous roots which may be red or white in colour; simple, ovate-cordate; flowers purple, corolla funnel-shaped; fruits capsules.

Flowering & Fruiting: MOctober-December.

Medicinal Parts: Tuberous roots.

Chemical constituents: Tubers – Hydrocyanic acid, oxalic acid, phytic acid, starch, vitamins, enzymes, β-sitosterol, oalmitic acid, NaCl.

Medicinal Properties: Roots – Aphrodisiac, diuretic, laxative, refrigerant, sweet.

Medicinal Uses: Roots Burning sensation, constipation, renal & Diabetes, vesical calculi, strangury.

229. *Ipomoea nil* (Linn.) Roth. (Convolvulaceae)

Syn.: *I. headeracea* auct. non (L.) Jacq.; *Convolvulus nil* L.; *C. bilobatus* Roxb.

Vern.: Kaladana.

Distribution: Common in mixed forests, on hedges and roadside bushes. Balaghat, Chhatarpur, Damoh, Dhar, Indore, Jabalpur, Panna, Sagar, Satna, Seoni, Tikamgarh.

Botany: Twining herbs, with ovate-cordate leaves and purple flowers in axillary cymes. Capsules 3-valved; seeds glabrous.

Flowering & Fruiting: September-March.

Medicinal Parts: Seeds.

Chemical constituents: Seeds contain – Alkaloids: Lysergol, chanoclavine, penniclavine, isopeniclavine, elymoclavine; Resin, Glucosides.

Medicinal Properties: Seeds – Acrid, anti-inflammatory, carminative. depurative, purgative, sweet.

Medicinal Uses: Seeds – Brochitis, constipation, fever, flatulence, gout, inflammation, leucoderma, scabies, skin diseases.

230. *Ipomoea sepiaria* Roxb. (Convolvulaceae)

Syn.: *I. maxima* auct. non (L. f.) Sweet.

Distribution: In wastelands & along drains. Balaghat, Damoh, Mandla.

Botany: Perennial twining plants, with Pink flowers and capsule fruits.

Flowering & Fruiting: August-January.

Medicinal Parts: Whole plant.

Medicinal Properties: Whole plant – Aphrodisiac, Cooling, Diuretic, Laxative, Rejuvenating, Sweet, Tonic.

Medicinal Uses: Whole plant – Burning sensation, General debility, Strangury.

231. *Ixora coccinea* L. (Rubiaceae)

Distribution: Chhatarpur. Planted.

Botany: A shrub. Leaves oval with cordate, amplexicaul base. Flowers deep Red, attractive, in Corymbs. Berries hard.

Flowering & Fruiting: February-April.

Medicinal Parts: Root, Leaves, Flowers.

Chemical Constituents: *Root* – Mannitol, Myristic acid, Cyanidin, Delphinidin, Methyl esters of Palmitic acid, Stearic acid, Oleic acid, Linoleic acid.

Flowers – Rutin, Cyanidin-3-rutinoside, Leucocynidine Glycoside.

Medicinal Properties: Root – Acrid, Antiseptic, Astingent, Febrifuge, Sedative, Stomachic.

Flowers – Astringent, Bitter, Carminative, Constipating, Digestive, Sweet.

Medicinal Uses: Root – Chronic ulcers, Diarrhoea, Dysentery, Fever, Gonorrhoea, Hiccough, Skin diseases, Sores.

Leaves – Diarrhoea.

Flowers – Dysentery, Dysmenorrhoea, Leucorrhoea, Sores, Ulcers.

232. *Jasminum arborescens* Roxb. (Oleaceae)

Syn.: *J. roxburghianum* Wall.

Vern.: Chameli.

Distribution: Common in mixed forests. Balaghat, Bhopal, Damoh, Hosahangabad, Mandla, Shivpuri.

Botany: A large shrub with hairy climbing branchlets; leaves simple, opposite, ovate-oblong; flowers White, Fragrant, panicled; fruits Berry.

Flowering & Fruiting: March-October.

Medicinal Parts: Leaves.

Chemical Constituents: Oil contains Benzyl benzoate, Benzyl alcohol, Eugenol, Farnesot, Bergamotene, Nerolidol, P-cresol, Benzyl acetate, Bergamontane, Geranoil, Indole, Linalool, Benzoic acid, Vanilioin.

Medicinal Properties: Leaves – Astringent, Bitter, Emetic, Refrigerant, Stomachic, Sweet, Tonic.

Medicinal Uses: Leaves – Colic, Cough, Faltulence, General debility, Menstrual disorders.

233. *Jasminum auriculatum* Vahl. (Oliaceae)

Vern.: Juhi.

Distribution: Near streams and also cultivated. Rewa.

Botany: A bushy or climbing shrub. Leaves with a pair of minute leaflets or simple. Flowers White, Star shaped.

Flowering & Fruiting: July-August.

Medicinal Parts: Root, Flowers.

Chemical Constituents: Indole, Benzyl benzoate, Methyl Anthranilate, Benzyl Alcohol.

Medicinal Properties: Flowers – Acrid, Astringent, Bitter, Depurative, Diuretic, Fragrant, Refrigerant, Sweet.

Medicinal Uses: Root – Skin disease.

Flowers – Burning sensation, Hyperdipsia, Ophthalmopathy, Strangury, Ulcers.

234. *Jasminum grandiflorum* L. (Oleaceae)

Syn.: *J. officinale* L.

Vern.: Chameli.

Distribution: As forest undergrowth and commonly planted in gardesn. Bhopal, Chhatarpur, Hosahangabad.

Botany: Scandent or climbing shrubs, with imparipinnate leaves and white flowers in lax panicles.

Flowering & Fruiting: July-August.

Medicinal Parts: Root, Leaves, Flowers.

Chemical constituents: Leaves – Salicylic acid.

Flowers – Essential oil in flowers contains lactone, pyridine & nicotinate derivatives.

Medicinal Properties: Root & Leaves – Acrid, anodyne, anthelmintic, antiseptic, aphrodisiac, astringent, bitter, depurative, diuretic, emollient.

Flowers – Ophthalmic, refrigerant.

Medicinal Uses: Root – Amenorrhoea, constipation, dysmenorrhoea, flatulence, leprosy, mental debility, paralysis, ringworm, skin diseases. sterility, strangury.

Leaves – Corns, dysmenorrhoea, leprosy, otalgia, skin diseases, strangury, ulcers, wounds.

Flowers – Dysmenorrhoea, leprosy, pruritus, skin diseases, strangury, ulcers.

235. *Jasminum multiflorum* (Burm. f.) Andr. (Oleaceae)

Syn.: *J. pubescens* Willd.; *J. hirsutum* Willd.

Vern.: Safed Chameli.

Distribution: Common along roadsides and often cultivated in gardens. Balaghat, Bhopal, Jabalpur, Sidhi.

Botany: A scandent or diffuse, densely pubescent shrub. Leaves simple, ovate thick, cordate. Flowers White slightly scented in terminal, many-flowered umbles. Calyx teeth 1 cm. or more in length.

Flowering & Fruiting: December-June.

Medicinal Parts: Leaves, Flowers.

Chemical Constituents: Leaves – Friedelin, Lupeol, Betulin, □-amyrin, Betulinic acid, Ursolic acid, Oelanolic acid, Jasmolactones, A,B,C & D, Multiforoside, Multiroside, Jasmutiside, Multiflorin.

Medicinal Properties: Flowers – Acrid, Alexipharmic, Bitter, Depurative, Digestive.

Laxative, Refrigerant, Sweet.

Medicinal Uses: Flowers – Cephalagia, Inflammation, Rheumatism.

236. Jasminm sambac (Linn.) Ait. (Oleaceae)

Vern.: Mogra.

Distribution: As forest undergrowth and commonly planted in gardens, Hosahangabad.

Botany: Scandent or climbing shrubs, with simple leaves and white flowers.

Flowering & Fruiting: April-October.

Medicinal Parts: Roots, Leaves, Flowers.

Chemical constituents: Flowers contain Pyridine, nicotinate derivatives.

Medicinal Properties: Flowers – Acrid, alexipharmic, bitter, ophthalmic, purgative, refrigerant.

Medicinal Uses: Leaves – Fever, skin diseases, ulcers.

Flowers – Fever, leprosy, hiccough, insanity, pruritus, ulcers, vomiting.

237. *Jatropha curcas* L. (Euphorbiaceae)

Vern.: Ratanjot.

Distribution: Common along cultivated fields and in wastelands. Balaghat, Chhatarpur, Indore, Mandla, Panna, Seoni, Shahdol, Sidhi, W. Nimar.

Botany: Eglandular, soft wooded shrubs or small trees, with 3 to 5-angled ovate-cordet leaves. Flowers greenish-yellow, in corymbose cymes. Capsules globose-ovoid.

Flowering & Fruiting: August-January.

Medicinal Parts: Root bark, leaves, stem, seeds, and latex.

Chemical constituents: Seed contains Amino acids, arachid acid, lionleic acid, myrstic acid, oleic acid, stearic acid, arabinose, glucose, rhamnose, galactose, xylose, galactouronic acid, β – sitoserol.

Stem bark contains B – Sitosterol, taraxerol.

Leaves contain Sucrose, vitexin, compesterol, triacontanol.

Medicinal Properties: Fruit & seed – Anthelmintic, puragative.

Root – Astringent, carminative, digestive, emetic.

Medicinal Uses: Fruit & seed – Abdominal complaint, anemia, biliousness, diseases of the heart, dysentery, fistula, urinary discharges.

Latex – Styptic

Root bark – Anaemia, biliousness, cough, dysentery, fistula, gum trouble, inflammation, rheumatism, swelling, tuberculosis, ulcer.

238. *Jatropha gossypifolia* L. (Euphorbiaceae)

Vern.: Lal – erenda.

Distribution: Common along roadsides and cultivated fields. Bhind, Chhatarpur, Damoh, Dhar, Mandla, Panna, Rewa, Satna.

Botany: Tall, erect, glandular hairy shrubs, with deep reddish-purple branches. Leaves orbicular to obovate. Palmately 3 to 5 lobed. Flowers blood-red. Capsules 3-lobed With oblong, strophiolate seeds.

Flowering & Fruiting: April-September.

Medicinal Parts: Bark, leaves, seeds.

Chemical constituents: Seed contains Alanine, arginine, cystine, glycinc, leucine, methionine, valline, linoleic acid, oleic acid, palitic acid.

Leaves contain Apigenin, vitexin, Isovitexin.

Bark contain Ceryl & myricyl alcohol, palmitone, β-sitosterol.

Root contain Jatrophalones A & B.

Plant contain Gossypibetelene.

Medicinal Properties: Plant – Emetic, purgative, stimulant.

Root – Astringent, carminative, digestive.

Medicinal Uses: Leaves – Carbuncles, eczema, Itching.

Latex & seed oil – Ulcer (Externally applied).

239. *Justicia gendarussa Burm.* f. (Acanthaceae)

Syn.: *J. vulgaris* Nees.

Vern.: Nili-Nargandi.

Distribution: Commonly planted as hedge in gardens, also found as escape in wastelands. Bhopal, Chhatarpur, Mandla.

Botany: An erect undershrub; leaves simple, linear-lanceolate or lanceolate; flowers white, spotted purple within, clustered in the interrupted spikes; fruits capsules.

Flowering & Fruiting: January-February.

Medicinal Parts: Root, Leaves.

Chemical constituents: Root – β-Sitosterol.

Leaves – β-Sitosterol, Alkaloid.

Medicinal Properties: Roots & Leaves – Acrid, anodyne, anti-inflammatory, antiperoidic, bitter, diaphoretic, emetic, expectorant, insecticidal.

Medicinal Uses: Roots & Leaves – Amenorrhoea, Arthritis, bronchitis, cough, dysmenorrhoea, fevers, otalgia, rheumatism.

240. *Justicia procumbens* L. (ACANTHACEAE)

Syn.: *Rostellularia procumbens* (L.) Nees.

Distribution: Common on rocky hill slopes. Balaghat, Hosahangabad, Indore.

Botany: A procumbent, diffuse, slender, branching annual.

Flowering & Fruiting: September-December.

Medicinal parts: All parts

Chemical constituents: *Leaves contain* – An essential oil which is Antifungal; Justicidins A & B (major), Diphyllin and Lignan Neojusticin.

Medicinal properties: Alterative, Anthelmintic, Diaphoretic, Diuretic, Expectorant, Febrifuge, Laxative.

Medicinal uses: Asthma, Cough, Backache, Flatulence, Lumbago.

241. *Kalanchoe pinnata* (Lam.) Pers. (Crassulaceae)

Vern.: Pattharchur.

Distribution: Planted and found as escape on border of forests. Balaghat, Chhatarpur, Mandla.

Botany: A perennial glabrous succulent herb. Leaves opposite-decussate, Compound, Leaflets ovate or elliptic, crenate or serrate. Flowers in large spreading panicles, pendent, pale-greenish and reddish-purple.

Flowering & Fruiting: October-February.

Medicinal Parts: Leaves and Its juice.

Chemical constituents: Leaves contain Alkanes, Tricontane, Alpha – & Beta-Amyrin, Beta-sitosterol, Fumaric acid, Mallic acid, Citric acid, Quercetin, Kaempferol, Calcium sulphate, Calcium oxalate.

Medicinal Properties: Leaves – Astringent, stimulant, styptic, wound healing.

Medicinal Uses: Leaves – Diarrhoea, disuria, dysentery, polyuria, tuberculosis, ulcers.

Leaves (Externally applied) – Bleeding, calculii, cuts, ulcers, wounds.

Ayurvedic preperations: *Parnabija svarasa.*

242. *Kyllinga nemoralis* (Forster) Dandy. ex Hutch. (Cyperaceae)

Syn.: *K. monocephala* Rottb.; *Cyperus kyllingia* Endl.

Distribution: Jabalpur, Ujjain.

Botany: An erect glabrous sedge with elongate creeping rhizome; flowers minute in spike, solitary or 2-3 togethere, the rachis naked when spikelets fall off; spikelets one-flowered; fruits nuts.

Flowering & Fruiting: August-October.

Medicinal Parts: Tubers.

Medicinal Properties: Tuber – Aromatic, astringent, bitter, demulcent, diuretic, expectorant, febrifuge, fragrant, sweet.

Medicinal Uses: Tuber – Bronchitis, cough, diabetes, diarrhoea, dysentery, fever, stangury, tumours.

243. *Lablab purpureus* L. (Fabaceae)

Syn.: *Dolichos lablab* L.

Vern.: Sem.

Distribution: Bhopal, Chhatarpur, Damoh, Hosahangabad, Jabalpur, Sagar, Sidhi.

Botany: A tall, glabrous twining perennial herb, with white or pink flowers & white pods.

Flowering & Fruiting: November-April.

Medicinal Parts: Leaves, Seeds.

Chemical constituents: Seeds – Hydrocarbons, proteins, lipids, amino acids, tryptophan, arginine, tyrosinase, hydrocyanic acid, calcium, phosphorus, iron, vitamins-A,B, B_2, C.

Pods – L-pipecolic acid, galactusonic acid, galactose, glucose & arabinose.

Medicinal Properties: Leaves – Alexipharmic.

Seed – Aphrodisiac, astringent, diuretic, laxative, sour, stomachic, sweet.

Medicinal Uses: Leaves – Colic.

Seed – Inflammations, stranugry.

244. *Lagenaria siceraria* (Mol.) Standley. (Cucurbitaceae)

Syn.: *L. leucantha* (Duchesne) Rusby.

Vern.: Lauki.

Distribution: Cultivated. Bhopal, Chhindwada, Mandla, Morena, Sidhi.

Botany: Climbing herbs, with angular or shortly 3-lobed, white flowers & bottle shaped fruits.

Flowering & Fruiting: July-February.

Medicinal Parts: Whole plant.

Chemical constituents: Fruits – Pectin, cucurbitacins, triterpenoids.

Seed – Rhamnose, glucose, fructose, galactose, sucrose, raffinose, stachyose.

Seed oil – Saponin, fatty oil, palmitic, palmitoleic, oleic, stearic & linoleic acids.

Medicinal Properties: Root – Anti-inflammatory, emetic, purgative.

Leaves – Anodyne, bitter, depurative, emetic, expectorant, febriuge, purgative, refrigerant.

Flowers – Cooling.

Fruits – Acrid, anti-inlammatory, bitter, depurative, diuretic, emetic, expectorant, febrifuge, purgative, refrigerant.

Seeds – Brain tonic, cooling, diuretic, purgative.

Medicinal Uses: Root – Constipation, Inflammations.

Leaves – Asthma, bronchitis, constipation, cough, fever, flatulance, inflammations, jaundice, leprosy, skin diseases.

Flowers – Useful in Odontalgia & Ophthalmia.

Fruits – Bronchitis, cough, fever, inflammations, leprosy, skin diseases.

Seed – Cough, dropsy, fever, inflammations.

245. *Lagerstroemia speciosa* (L.) Pers. (Lythraceae)

Syn.: *L. flosreginae* Retz.

Distribution: Balaghat, Hosahangabad.

Botany: A deciduous tree, with large showy flowers and woody capsule fruits.

Flowering & Fruiting: June-January.

Medicinal Parts: Fruits, Seeds, Leaves, Bark, Root.

Chemical Constituents: Leaves – Alanine, Isoleucine, a-aminobutyric acid, Menthoeonine, Lageracetal, Amylalcohol, Ellagic acid, β-sitosterol, lager tannin.

Fruit, Bark – Tannin.

Medicinal Properties: Root – Astringent, Febrifug, Stimulant.

Bark / Leaves – Purgative.

Seeds – Narcotic.

Medicinal Uses: Bark – Fever.

246. *Lannea coromandelica Merr.* (Anacardiaceae)

Syn.: *L. grandis* Engl.

Distribution: Common in dry deciduous forests. Bhopal, Hosahangabad, Indore, Jabalpur, Mandla, Seoni, Shivpuri.

Botany: Trees, leaves pinnately compound; flowers Purplish; ripe fruits red drupes.

Flowering & Fruiting: *February-May.*

Medicinal parts: Bark, Leaves, Gum

Chemical constituents: *Bark contains*-β-Sitosterol, Physcion, Phlobatannins, Physcion anthranol-B.

Leaves contain – Ellagic acid, Quercetin, β-Sitosterol, Rutin, Leucocyandin, Leucodelphinidin.

Flowers contain – Quercetin, Iso-quercitrin, Morin, Ellagic acid.

Medicinal properties: *Bark* – Acrid, Astringent, Anodyne, Anticancer, Antibacterial, Stomachic, Sweet, Thermogenic.

Leaves – Anticancer, Anti inflammation, Neuralgic.

Medicinal uses: *Bark* – Cuts, Diarrhoea, Dysentery, Gout, Wounds, Heart complaints, Opthalmia, Stomatitis, Ulcers.

Leaves – Bodyache, Elephantitis, Inflammation, Swellings.

247. *Lantana camara* L. (Verbenaceae)

Distribution: On roadsides and forest margins. Balaghat, Chhatarpur, Damoh, Dhar, Hosahangabad, Indore, Narsimhapur, Panna, Satna, Shivpuri, Sidhi, Tikamgarh.

Botany: Prickly shrubs, prickles hooked. Leaves ovate or ovate-oblong. Flowers multicoloured, orange-yellow, pink or red. Drupes globose, black when ripe.

Flowering & Fruiting: Throughout the Year.

Medicinal Parts: Whole plant.

Chemical constituents: Plant contains Essential oil.

Leaves contain Lantanine, lantadene, arachidic acid, linoleic acid, myristic acid, palmitic acid, citral, eugenol, furfural, geraniol, linalool, Alpha – amyrin, Beta – sitosterol. Lantanin acid, lantadene – A & B. verbascoses & lantanane.

Medicinal Properties: Antiseptic, antispasmodic, carminative, diaphoretic, laxative.

Medicinal Uses: Fistula, malaria, rheumatim, wounds.

248. *Lawsonia inermis* L. (Lythraceae)

Syn.: *L. alba* Lamk. Egyptian

Vern.: Mehndi.

Distribution: Almost throughout the State; usually planted as hedge.

Botany: Much branched shrubs, with broadly lanceolate leaves and creamish, fragrant flowers. Capsules globose.

Flowering & Fruiting: May – November.

Medicinal Parts: Leaves, flowers, fruits.

Chemical constituents: Leaves contain Heena dye, Hanno – tannic acid, tannin, olive green resin.

Medicinal Properties: Anti – inflammatory, antipyretic, burning sensation, diuretic, emetic, expectorant, prophylactic.

Medicinal Uses: Amenorrhoea, bronchitis, cough, haemorrhage, headache, leprosy, leucoderma, lumbago, scabies, skin diseases, stomatitis, ulcer.

Ayurvedic preperations: *Madyantyadi churna, Mahanila ghrit.*

249. *Lens culinaris Med.*(Fabaceae)

Vern.: Masur.

Distribution: Cultivated for its seeds, cooked as a pulse. Chhatarpur, Hosahangabad, Mandla, Sagar, Shivpuri.

Botany: A small erect soft pubescent herb, with white Rose, Red or Violet flowers & pod fruits.

Flowering & Fruiting: November-April.

Medicinal Parts: Seeds.

Chemical constituents: Seeds contain – Proteins, vitamin-B, tricetin, luteolin, kaempferol glycoside, 3, 4, 7 – trihydroxyflavone, proanthocyanidins, diglycosyldelphinidin.

Chemical constituents: Seeds – Astringent, constipating, diuretic, sweet.

Medicinal Uses: Seeds – Anaemia, diarrhoea, dysentery, skin diseases, strangury, tumours.

250. *Lepidium sativum* L. (Brassicaceae)

Vern.: Halim.

Distribution: Cultivated. Sidhi.

Botany: Erect annual herbs, with pinnatifid leaves & white flowers fruits pod.

Flowering & Fruiting: February-May.

Medicinal Parts: Roots, Leaves, Seeds.

Chemical constituents: Glucotropoeoline, sinapine, sinapic acid, β-sitosterol, benzylcyanide, vitamins, essential oil.

Medicinal Properties: Root – Acrid, bitter.

Leaves – Antibacterial, diuretic, stimulant.

Seeds – Aphrodisiac, bitter, depurative, diuretic, rubefacient.

Medicinal Uses: Seeds – Asthma, cough, diarrhoea, dysentery, dyspepsia, hiccough, leucorrhoea, lumbago, scurvy, seminal weakness, skin diseases.

251. *Leptadenia reticulata* Wight & Arn. (Aaclepiadaceae)

Distribution: In Sal forests, mostly on red morrum or sandy soil. Hoshangabad.

Botany: Perennial climbers, with green or yellow flowers and follicle fruits.

Flowering & Fruiting: July-April.

Medicinal Parts: Root.

Medicinal Properties: Stimulant, Restorative.

Medicinal Uses: Leaves/Roots – Skin affections.

Plant – Habitual abortion.

252. *Leucas cephalotes* (Roth.) Spreng. (Lamiaceae)

Syn.: *L. aspera* (Willd.) Link.

Vern.: Dhurpi sag.

Distribution: Amidst grasses in forest clearings and in cultivated fields. Balaghat, Betul, Chhatarpur, Damoh, Dhar, Sagar, Satna, Seoni, Sidhi, Tikamgarh, W. Nimar.

Botany: Erect, hispid herbs, with ovate, crenate-serrate leaves. Flowers white, in large, globose terminal heads. Calyx-tube bent, mouth produced on the upper side, villous within. Nutlets obovoid-oblong.

Flowering & Fruiting: August-April.

Medicinal Parts: Whole plant, flowers.

Chemical constituents: Plant contains B – Sitosterol, flavonoides, glycosides, glucose, alkaloids, baballenic acid, aromatic oil.

Medicinal Properties: Anthelmintic, antiseptic, antispasmodic, appetiser, carminative, cathartic, diaphoretic, expectorant, heating, indigestible, laxative, stimulant.

Medicinal Uses: Asthma, bronchitis, cold, cough, dyspepsia, eruptions, eye complaint, headache, inflammation, jaundice, leucoma, paralysis, scabies, skin diseases, snake bite, swelling.

Ayurvedic preperations: *Vishmajavaradi kshar, Sudarshan churna.*

253. *Limnophilla aromatica* (L.) Merr. (Scrophulariaceae)

Syn.: *L. gratissima* Blume.

Vern.: *Kuttra.*

Distribution: In moist habitats, particularly in cultivated fields. Balaghat, Damoh, Hosahangabad.

Botany: An aromatic herb, with Purplish flowers and capsule fruit. *Flowering & Fruiting:* November-February.

Medicinal Parts: Whole plant.

Chemical Constituents: The plant contains d-limonene, d-perillaldehyde, a flavone as 7-desmethyl artemetin.

Medicinal Properties: Whole plant – Anthlemintic, Anti-inflammatory, Antiseptic, Aperient, Appetiser, Carminative, Digestive, Diuretic, Emollient, Febrifuge, Refrigerant, Slightly bitter, Sour.

Medicinal Uses: Whole plant – Agalactia, Anorexia, Constipation, Dyspepsia, Fever, Helminthiasis, Inflammation, Strangury.

254. *Limonia acidissima* L. (Rutaceae)

Syn.: *Feronia elephantum* Corr.; *F. limonia* (L.) Swing.

Vern.: Kabit.

Distribution: Hoshangabad, Panna.

Botany: Medium sized trees, armed spine. Leaves imparipinnate, with 3-9 leaflets. Flowers pentamerous, yellowish. Fruit globose, with woody rind.

Flowering & Fruiting: February– October.

Medicinal Parts: Leaves, fruits & gum.

Chemical constituents: Fruits contain Riboflavine, vitamin – C pectin, Iron, phosphorus.

Leaves contain Volatile oil. (Estragol).

Medicinal Properties: Fruit – Acrid, alexipharmic, aphrodisiac, refrigerant, sour, sweet.

Leaf – Astringent, carminative.

Medicinal Uses: Fruits – Asthma, biliousness, blood diseases, cough, dysentery, earache, heart diseases, hiccough, leucorrhoea, ophthalmia, throat diseases, tumour, vomiting.

Seed – Headache, heart diseases.

Leaf – Dysentery, hiccough, vomiting.

255. *Linum usitatissimum* L. (Linaceae)

Vern.: Alsi.

Distribution: Cultivated. Hosahangabad, Sidhi.

Botany: An annual herb; leaves linear-lanceolate; flowers bluish violet or white in terminal panicles; fruit capsule.

Flowering & Fruiting: October-December.

Medicinal Parts: Flowers, Seeds, Seed oil.

Chemical constituents: Seed – Proteins, glucosides, phytin, lecithin, β-carotenoids, cholesterol, campesterol, stigmasterol, sitosterol, 5-dehydro-avenasterol, cyclosterol & 24-methylenecycloartanol.

Seed oil – Proteins, wax, resin, sugar, phosphates, linamarin.

Medicinal Properties: Flowers – Cardiotonic.

Seeds – Acrid, bitter, diuretic, emollient, expectorant, laxative, sweet, thermogenic.

Medicinal Uses: Seed – Crushed linseed is used as poultice for relief of Boils & carbuncles, local inflammations, ulcers.

Seed oil – Linseed oil is used for external application.

256. *Ludwigia octovalvis Raven.* (Onagraceae)

Syn.: *Jussiaea suffruticosa* L.; *Sessiliflora raven* Michl.

Distribution: Common near streams, cultivated fields, river and ponds. Balaghat, Hosahangabad, Jabalpur, Mandla, Narsimhapur, Satna, Sagar, Seoni, Shahdol, Sidhi.

Botany: Much branched perennial under shrubs; with axillary, large, solitary, yellow flowers & 8-ribbed capsule fruits.

Flowering & Fruiting: *Throughout the year.*

Medicinal parts: Whole plant

Chemical constituents: Orientin, Isoorientin, Vitexin, Isovitexin.

Medicinal properties: Astringent, Anti-inflammatory, Carminative, Diuretic, Expectorant, Febrifuge, Laxative, Vermifuge.

Medicinal uses: Asthma, Cough, Dropsy, Diarrhoea, Dyspepsia, Dysentery, Fever, Flatulence, Leucorrhoea, Strangury.

257. *Luffa acutangula* (L.) Roxb. var. *amara* (Roxb.) C. B. Clarke (Cucurbitaceae)

Vern.: Kadvi turoi.

Distribution: Bhopal, Sidhi.

Botany: Extensive climbers, with 3-fid tendrils and palmetely 5 to 7 – angled or lobed leaves. Flowers yellow; stamens 3. Fruits 10-angled.

Flowering & Fruiting: June – October.

Medicinal Parts: Fruits, Seeds, Stem.

Chemical constituents: Fruits contain Bitter alkaloids

Seed contains Oil & Albumin

Medicinal Properties: Plant – Acrid, alexiteric, bitter, carminative, digestive, laxative, Pungent.

Fruits – Cathartic, emetic.

Medicinal Uses: Plant – Anaemia, ascites, biliousness, bronchitis, inflammation, jaundice, leucoderma, liver complaint, piles, tumor, uterine & vaginal tumours.

Fruits – Asthma, biliousness, bronchitis, fever, headache, leucoderma, urinary discharges.

Seed – Headache.

258. *Luffa cylindrica* (L.) M. Roem. (Cucurbitaceae)

Syn.: *L. aegyptiaca* Mill.

Vern.: Ghia turoi.

Distribution: Bhopal, Sidhi. Cultivated.

Botany: A large monoecious climber with stout 5-angled twisted stems; Male flowers in axillary racemes, crowded at the top, female flowers solitary; fruits large, cylindric, smooth, 10-ribbed; seeds black or grey.

Flowering & Fruiting: June-December.

Medicinal Parts: Fruits, Seeds.

Chemical constituents: Colosynthine, luffein.

Medicinal Porperties: Fruits – Anthlemintic, carminative, depurative, diuretic, emollient, expectorant, laxative.

Medicinal Uses: Fruits – Bronchitis, fever, leprosy, syphilis, tumours.

259. *Macrotyloma uniflorum* (Lam.) Verdc. (Fabaceae)

Syn.: *Dolichos biflorus* L.

Distribution: On roadside shrubs. Hosahangabad.

Botany: Trailing annual plants, with Yellow flowers & Pods fruits. *Flowering & Fruiting:* September-November.

Medicinal Parts: Seeds.

Chemical Constituents: Seed contains – Urease, Strepogenin, β-sitosterol, Genistein, Kievitone, Phaseollidin, Isoferreirin, Coumesterol, Psoralidin, d-glucose, d-galactose, L-rhamnose, d-arabinose, L-ascorbic acid, Amino acids-glycine, Alanine, Cysteine, Serine, Aspartic acid.

Medicinal Properties: Seed – Acrid, Anthlemintic, Astringent, Bitter, Diaphoretic, Diuretic, Expectorant, Febrifuge, Ophthalmic, Thermogenic, Tonic.

Medicinal Uses: Seed – Asthma, Bronchitis, Hiccough, Tumours, Strangury, Verminosis.

260. *Madhuca indica* Gmel. (Sapotaceae)

Syn.: *M. longifolia* (Koenig) Mach. Var. *latifolia* Chev.; *Bassia latifolia* Roxb.

Vern.: Mahua.

Distribution: Common throughout the state on roadsides, near villages and in dry mixed deciduous forests. Chhatarpur, Damoh, Shivpuri, Sidhi.

Botany: Large, deciduous trees, with elliptic-obovate leaves and white flowers in axillary fascicles. Berries ovoid.

Flowering & Fruiting: April – June.

Medicinal Parts: Flower, Fruit, Leaves, Bark, Seed oil.

Chemical constituents: Flowers Sugar, cellulose.

Seeds Oil.

Medicinal Properties: Plant – Acrid, anthemintic, aphrodisiac, bitter, cooling, flattening, sweet.

Flowers – Aphrodisiac, cooling, sweet.

Fruits – Aphrodisiac, cooling, indigestive, sweet.

Seed oil – Acrid, sweet.

Medicinal Uses: Plant – Biliousness, burning sensation, ulcer.

Bark – Bone fracture, leprosy, ulcer, wounds.

Flower – Biliousness, burning sensation, ear complaint, heart diseases.

Fruits – blood diseases, bronchitis, heart diseases.

Seed oil – Burning sensation.

Ayurvedic preperations: *Madhukasav, Kutajarishta, Kanakasava, Parthadyarishta, Madhukadi him.*

261. *Malaxis acuminata* D. Don. (Orchidaceae)

Syn.: *Microstylis wallichii* Lindl.

Distribution: In moist shady places on rocks, Hoshangabad.

Botany: A short stemmed terrestrial herb, with more or less pseudobulb at the base and fibrous roots; leaves simple, sheating, ovate or lanceolate; flowers minute pale Yellow-Green tinged with Purple in terminal racemes.

Flowering & Fruiting: July-February.

Medicinal Parts: Green swollen stem base covered by brown scales.

Chemical constituents: 3'-O-Methylbatatasin-III, β-sitosterol, sugars.

Medicinal Properties: Stem base – Aphrodisiac, febrifuge, refrigerant, sweet.

Medicinal Uses: Stem base – Burning sensation, fever, seminal weakness, tuberculosis.

262. *Mallotus philippensis* (Lam.) Muell.-Arg. (Euphorbiaceae)

Syn.: *Croton philippense* Lamk.; *Rottlera tinctoria* Roxb.

Distribution: Common in mixed forests along roadsides and river banks. Balaghat, Chhatarpur, Damoh, Hosahangabad, Panna, Satna, Seoni, Shahdol, W. Nimar.

Botany: Small, dioecious trees, with ovate to ovate-lanceolate leaves palmately veined and red gland-dotted beneath. Capsules globose.

Flowering & Fruiting: October-May.

Medicinal Parts: The glandular hairs of the fruit.

Chemical constituents: Seeds – Rottlerin, isorottlerin, corotoxigenin, coroglaucigenin.

Heartwood – Betulin-3-acetate, lupeol, lupeol acetate, sitosterol, bergenin.

Bark – Acetylaleuritolic acid, α-amyrin, sitosterol, bergenin.

Medicinal Properties: Glandular hair – Acrid, alexipharmic, depurative, digestive, purgative, styptic, thermogenic, vermifuge.

Medicinal Uses: Glandular hair – Constipation, cough, flatulence, haemorrhages, ringworm, scabies, ulcers, wounds.

263. Mangifera indica L. (Anacardiaceae)

Vern.: Aam.

Distribution: Commonly planted throughout the State. Bhopal, Chhindwada, Damoh, Hosahangabad, Jabalpur, Mandla, Raisen, Rewa, Shahdol, Shivpuri.

Botany: Evergreen trees, with oblong-lanceolate leaves and polygamous flowers in panicles. Stamens and style solitary. Drupes yellow when ripe.

Flowering & Fruiting: Febuary – June.

Medicinal Parts: Leaves, bark, flower, fruits, seed, gum.

Chemical constituents: Mangiferin, Amino acids, Gallotanin, Gallic acid, Ethylgallati, Isoquercetin, β – Sitosterol, Riboflavin, Vitamin – C, citronellal, Deterpene, Geraniol, Limonene, mangiferol, Mangiferon, Nerol & Tannine, Vitamin – A, Ascorbic acid, protein, fat, carbohydrate, Calcium, phosphorous, Iron, fructose, sucrose, glucose.

Medicinal Properties: Bark – Acrid, astringent, cooling.

Leaf – Acrid, astringent.

Flower – Appetizer, astringent, cooling.

Unripe fruits – Acrid, astringent, sour.

Ripe fruits – aphrodisiac, appetiser, astringent, cooling, sweet.

Seed – Acrid, sweet.

Seed oil – Acrid, bitter, sweet.

Medicinal Uses: Unripe fruits – Biliousness, blood impurities, dysentery, throat trouble, ulcer, urinary dischargers, vaginal trouble.

Flower – Biliousness, blood diseases, bronchitis, dysentery, leucorrhoea, urinary discharges.

Ripe fruit – Blood impurities, heart trouble, ulcers, urinary discharges.

Seed – Dysentery, vomiting.

Seed oil – Stomatitis.

Ayurvedic preperations: *Pusyanug churna, Amrapanak and Amra murabba.*

264. *Manihot esculenta* Crantz. (Euphorbiaceae)

Syn.: *M. utilissima* Pohl.; *M. palmata* Muell.-Arg.; *M. aipi* Phol.; *M. dulcis* Pax.

Vern.: Cassava.

Distribution: Commonly grown in gardens for its edible tuberous roots, Indore.

Botany: A shrub, with palmately compound leaves and Yellow or Greenish White unisexual flowers.

Flowering & Fruiting: February-March.

Medicinal Parts: Tuberous roots.

Chemical Constituents: Plant contains Cyanogenetic Glucose. Juice is poisonous.

Medicinal Properties: Tuberous roots – Aperient, Appetiser, Nourishing, Sweet, Tonic, Vulnerary.

Medicinal Uses: Tuberous roots – Anorexia, Constipation, Dyspepsia, Foul ulcers, General debility, Wounds.

265. *Martynia annua* L. (Martyniaceae)

Vern.: Bicchu.

Distribution: In wastelands. Balaghat, Damoh, Dhar, Hosahangabad, Indore, Mandla, Panna, Ratlam, Rewa, Sagar, Satna, Sidhi.

Botany: Glandular pubescent herbs, with fistular stem, ovate-cordate leaves & drooping, pinkish-white flowers. Fruits woody, beetle-shaped, hooked at the tip.

Flowering & Fruiting: August-November.

Medicinal Parts: Fruit, Leaves.

Chemical Constituents: Leaves – Chlorogenic acid.

Seeds – Pale Yellow semi drying oil, Fatty oil: Fatty acids of the fatty oil are Palmitic, Stearic, arachidic, Oleic acid, Linoleic etc.

Fruits – Fatty oil: Fatty acids of the fatty oil are Plamitic, Stearic, Oleic acid, Linoleic acid.

Medicinal Properties: Fruits – Alexiteric, Anti-Inflammatory.

Medicinal Uses: Fruits – Inflammation.

Leaves – Epilepsy, Sore throat.

266. *Melia azedarach* L. (Meliaceae)

Vern.: Bakayan.

Distribution: Usually planted throughout Madhya Pradesh.

Botany: Trees, with 1 to 3-pinnate leaves and liliac flowers. Drupes 5-located, each locule 1-seeded.

Flowering & Fruiting: April – May.

Medicinal Parts: Root, leaf, seed, flowers.

Chemical constituents: Plant contains Azaridine, Bakayanin, Bakalactone, Morgosine, Azadirone, Azadiradione, Orchinol, Ochinine, Ochinolal, Nimbolinin, Nimbolidins, Triterpene, Meliacins, Sendanins, Melianol, Melianone, Lupeol, β – stiosterol, Cataechin,

Vanillin, Cinnamic acid, Cystine, Serin, Arginine, Glycine, Glutomic acid, Leucine, Threonine, Methionine.

Medicinal Properties: Root – Acrid, anodyne, anthelmintic, antiseptic, astringent, bitter, constpating, depurative, expectorant.

Seed – Anthelmintic, aphrodisiac, bitter, expectorant.

Flowers – Anodyne, astringent, diuratic, stomachic.

Medicinal Uses: Root – Amenorrhoea, asthma, burning sensation, cough, diabetes, dysenetery, fever, general debility helmintic, leprosy, leucoderma, lumbago, sciatica, skin diseases, ulcer, vomit, wound.

Leaf – Amenorrhoea, bronchitis, cardiac diseases, cough, hysteria, inflammation, leprosy, scabies.

Seed – Fever, helmintic, typhoid.

Flower – Dysenetery, fever.

Ayurvedic preperations: *Arshoghni vati.*

267. *Mentha spicata* L. (Lamiaceae)

Syn.: *Mentha Vidrdis* L.

Vern.: Podina.

Distribution: Cultivated throughout the state as field crops.

Botany: Aromatic, creeping or ascending herbs. Leaves ovate or ovate-elliptic. Flowers purplish.

Flowering & Fruiting: June-July.

Medicinal Parts: Leaves.

Chemical Constituents: Diosmetin, Diosmin from leaves. Essential oil contains Carvacrol, L-carvone, Dihydrocarveol, Limonene, Dipentene and Esters.

Medicinal Properties: Antispasmodic, Carminative, Stimulant.

Medicinal Uses: Leaves – Bronchtis, Fever.

268. *Merremia emarginata* (Burm. f.) Hall. f. (Convolvulaceae)

Syn.: *M. gangetica* (L.) Cuf.; *Ipomoea reniformis* Choisy.

Distribution: In wastelands, along edges of fields and on sandy slopes. Bhopal, Chhatarpur, Damoh, Seoni.

Botany: A creeping perennial herb rooting at the nodes; Leaves simple, long stalked, reniform or ovate-cordate, flowers yellow, axillary on 1-3 flowered, very short peduncles; fruits subglbose capsules with 2-4 glabrous light brown seeds.

Flowering & Fruiting: December-March.

Medicinal Parts: Whole plant.

Chemical constituents: Seeds contain Caffic acid, ferrulic acid, sinapic acid, p-coumaric acid.

Medicinal Properties: Plant – Acrid, anthelmintic, bitter, carminative, digestive, diuretic, refrigerant.

Medicinal Uses: Plant – Anaemia, cardiac diseases, fever, leucoderma, strangury.

269. *Merremia tridentata* (L.) Hallier f ssp. *Tridentata* (Convolvulaceae)

Syn.: *Ipomoea tridentata* (L.) Roth.

Distribution: In wastelands & river beds.

Botany: A small glabrous perennial herb with many decumbent stems from a small woody rootstock. Leaves sessile. Flowers axillary, peduncles wiry, 1-3-flowered; pedicels, sub-clavate. Sepals elliptic, glabrous. Capsules ovoid, glabrous, 2-celled. Seeds dark, glabrous.

Flowering & Fruiting: September-December.

Medicinal Parts: Whole plant.

Medicinal Properties: Anodyne, Aphrodisiac, Astringent, Bitter, Laxative, Thermogenic, Tonic.

Medicinal Uses: Plant – General debility, Haemorrhoids, Inflammation, Inflammation, Uropathy.

270. *Michelia champaca* L. (Magnoliaceae)

Vern.: Champa.

Distribution: Planted in gardens as ornamental. Hoshangabad.

Botany: A tall handsome evergreen tree, with straight trunk. Flowers solitary axillary enclosed in bud by a greyish-yellow pubescent, spathaceous deciduous bract. Fruit capsule.

Flowering & Fruiting: April-July.

Medicinal Parts: Whole plant.

Chemical constituents: Bark & Flowers contain – Essential oil contains: Mono – & sequiterpenes.

Medicinal Properties: Root / Root bark – Purgative.

Stem bark – Astringent, diuretic, expectorant, febrifuge, stimulant.

Flowers/ Floral bud/ Fruits – Acrid, anthelmintic, antipyretic, astringent, bitter, carminative, depurative, digestive, diuretic, expectorant, refrigerant, stomachic.

Medicinal Uses: Root/ Root bark – Amenorrhoea, constipation, dysmenorrhoea, inflammations.

Stem bark – Bronchitis, cough, debility, fever, stangury.

Flowers/ Floral bud/ Fruits – Amenorrheoa, bronchitis, burning sensation, colic, cough, dysmenorrhoea, dyspepsia, fever, flatulence, gout, leprosy, pruritus, skin diseases, ulcers, wounds.

271. *Mimosa pudica* L. (Mimosaceae)

Vern.: Chuimui.

Distribution: In forest openings. Balaghat, Bhopal, Damoh, Mandla, Seoni, Shivpuri.

Botany: A diffuse undershrub clothed with long bristles from bulbous bases. Leaves sensitive, digitate; pinnae 1-2 pairs, leaflets 12-20 pairs. Flower heads dense, globose, long peduncled, very small. Corolla pink or purple. Stamens 4. Pods flat, slightly curved, with 3-5 segmented valves; sutures prickly and persistent.

Flowering & Fruiting: August-March.

Medicinal Parts: Root, leaf.

Chemical constituents: Leaves contain Stigma sterol, Leucoanthocyanidin, Flavon, Glycoside, Alkaloid – Mimosine.

Seed contains Mucilage & free Amino acids.

Seed oil contains Linolic acid, Oleic acid, Palmitic acid, Stearic acid & Behenic acid.

Medicinal Properties: Root – Acrid, antispasmodic, astringent, bitter, constipating, diuretic, emetic, febrifuge.

Leaf – Bitter, Sudorfic.

Medicinal Uses: Root – Asthma, burning sensation, dyspepsia, fever, fistula, inflammation, jaundice, leucoderma, small pox, ulcer.

Leaf – Conjunctivitis, fistula, hemorrhage hydrocele, scrofula, wounds.

Ayurvedic preperations: *Lajjalu svarasa, Visagarbha taila.*

272. *Mimusops elengi* L. (Sapotaceae)

Vern.: Molsari.

Distribution: Usually planted in gardens and parks. Bhopal, Chhatarpur, Damoh.

Botany: Evergreen trees, with elliptic-oblong leaves. Flowers white with pinkish tinge, axillary, solitary or fascicled; corolla 8-lobed; stamens 8. Berries elliptic-oblong.

Flowering & Fruiting: January-September.

Medicinal Parts: Bark, flowers, fruits, seed.

Chemical constituents: Bark contains Lupeol & Xylose.

Leaves contain Myrcetin, Quercetin.

Flower contains Glucosides, Quercitol, Ursolic acid, Glucose, Triterpene, Essential oil, alcohol.

Seed contains Fatty oils, fatty acids – Palmitic acid, Stearic acid, Behenic acid, Oleic acid, Linoleic acid, Lauric acid, Myristic acid, Saponin.

Fruit contains Sugars.

Medicinal Properties: Bark – Acrid, anthelmintic, astringent, cooling.

Flower – Acrid, anthelmintic, astringent, cooling.

Fruits – Acrid, anthelmintic, astringent, cooling.

Seed – Acrid, anthelmintic, astringent, cooling.

Medicinal Uses: Bark – Dental troubles, diarrhoea, dysentery, pyorrhoea, stomatitis.

Flower – Cough, headache, leucoderma, skeletol fracture, worms.

273. *Mimusops hexandra* (Roxb.) Dub. (Sapotaceae)

Syn.: *Manilkara hexandra* (Roxb.) Dub.

Vern.: Khirni.

Distribution: Along roadsides and near strems. Chhatarpur, Damoh, Dhar, Hosahangabad, Indore, Panna, Raisen, Rewa, Sagar, Satna.

Botany: Small trees; bark rough, grey; Leaves oblong, elliptic or elliptic-ovate; flowers white, in fascicles of 2-6 and from the axils of follen leaves; fruits oblond-ellipsoid, reddish-yellow when ripe; seeds ovoid, reddish-brown, shining.

Flowering & Fruiting: September-December.

Medicinal Parts: Fruits.

Medicinal Properties: Fruit – Aphrodisiac, cooling, indigestion, sweet, tonic.

Medicinal Uses: Fruit – Biliousness, hallucinations, heart diseases, leprosy, loss of consciousness.

274. *Mirabilis jalapa* L. (Nyctaginaceae)

Vern.: Gulabbas.

Distribution: Cultivated in gardens and also as an escape in wastelands. Chhatarpur, Indore, Mandla, Tikamgarh.

Botany: Erect herbs, with tuberous roots and ovate leaves. Flowers scarlet, showy, in fascicles. Fruits subglobose.

Flowering & Fruiting: Throughout the year.

Medicinal Parts: Tuber, Root, Leaves.

Chemical constituents: Leaves – Alkaloid trigonelline, tricosan-12-one, n-hexacosanol, β-sitosterol, tetracosanoic acid; tarteric acid, citric acid, valine, tryptophan, leucine, alanine, glycine.

Medicinal Properties: Aphrodisiac, blood purifier, cooling.

Medicinal Uses: Blood purifier, goiter, piles.

275. *Mitragyna parvifolia* (Roxb.) Korth. (Rubiaceae)

Syn.: *Stephegyne parvifolia* Korth.

Vern.: Karam, Mundi.

Distribution: Balaghat, Bhopal, Chhatarpur, Damoh, Hosahangabad, Indore, Jabalpur, Mandla, Morena, Narsimhapur, Panna, Rewa, Sagar, Satna, Seoni, Shahdol, Shivpuri, Sidhi, Tikamgarh.

Botany: Trees, with panduriform or ovate to suborbicular leaves. Flowers in heads, greenish-yellow, with mitriform stigma. Capsules oblong-ellipsoid, 10-ribbed.

Flowering & Fruiting: August– January.

Medicinal Parts: Root, bark, leaves.

Medicinal Properties: Root & bark – Acrid, bitter, febrifuge, stomachic.

Leaf – Acrid, anodyne, anti – inflammatory, bitter, depurative, febrifuge, stomachic, styptic, sweet, vulnerary.

Medicinal Uses: Root & bark – Colic, fever.

Leaf – Colic, dyspepsia, erysipelas, fever, flatulence, inflammation, leprosy, skin diseases, wound.

276. *Momordica charantia* L. (Cucurbitaceae)

Vern.: Karela.

Distribution: Commonly cultivated throughout the State.

Botany: Monoecious climbers, with 5 to 7-lobed leaves and yellow flowers. Bracts of male flowers born about or below the middle of peduncle. Fruits muricate-tuberculate.

Flowering & Fruiting: July – October.

Medicinal Parts: Whole plant.

Chemical constituents: Plant contains Aromatic essential oil, Carotene, Glucoside, saponine, Momoridicine alkaloid, charantin (non nitrogenous substance).

Medicinal Properties: Root – Acrid, astringent, bitter.

Leaf – Antipyretic, bitter, emetic, purgataive.

Fruits – Acrid, anthelmintic, antidiabetic, anti – inflammatory, appetiser, bitter, depurative, digestive, febrifuge, purgative, stimulant, stomachic, thermogenic.

Medicinal Uses: Leaf – Burning sensation, constipation, fever, helminthiasis.

Fruits – Asthma, burning sensation, colic, constipation, cough, diabetes, fever gout, helminthiases, inflammation, leprosy, skin diseases, ulcer, wound.

Seed – Obstructions of the liver & spleen, ulcer.

277. *Momordica dioica* Roxb. ex Willd. (Cucurbitaceae)

Vern.: Jangli Karela.

Distribution: Commonly throughout the State.

Botany: Perinnial, dioecious climbers, with tuberous roots. Leaves ovate, entire or 3 to 5-lobed. Flowers yellow. Fruits ovoid, beaked, densely echinate with soft spines.

Flowering & Fruiting: June – October.

Medicinal Parts: Leaves, fruits, tuber.

Chemical constituents: Plant contains Alkaloids, mangenese.

Medicinal Properties: Leaf – Anthelmintic, aphrodisiac.

Fruits – Alexiteric, bitter, laxative, pungent, stomachic.

Medicinal Uses: Root – Headache, jaundice, urinary calculi.

Leaf – Asthma, bronchitis, fever, hiccough, piles.

Fruits – Asthma, biliousness, bronchitis, fever, heart trouble, leprosy, tumour, urinary discharges.

278. *Morinda pubescens* J.E. Smith. (Rubiaceae)

Syn.: *M. coreia* Buch-Ham.; *M. Tinctoria* Roxb.

Distribution: Balaghat, Indore, Morena, Satna, Seoni.

Botany: Moderate sized deciduous trees, with White flowers & drupe fruits.

Flowering & Fruiting: March-August.

Medicinal Parts: Roots, Leaves.

Chemical Constituents: *Root* – Morindone and its Glucoside Morindine; Damnacanthal, Nordamnacanthal, β-Sitosterol, Anthragallol-2-3-dimethyl ether, Soranjidol, Ibericin, Morindone-6-Primeveroside.

Leaves – Campesterol, Stigmasterol, β-sitosterol, Ursolic acid.

Medicinal Properties: Root – Styptic, Constipating, Anti-inflammatory, Alexeteric, Tonic.

Leaves – Digestive, Carminative, Febrifuge, tonic.

Medicinal Uses: Root – Boils, Dysentery, General debility, Haemorrhages, Inflammation.

Leaves – Diarrhoea, Dyspepsia, Gaotropathy, Gout, Inflammation,

Ulcerative stomatitis, Wounds.

279. *Moringa oleifera Lam.* (Moringaceae)

Syn.: *M. pterygosperma* Gaertn.

Vern.: Munga, Sahjan.

Distribution: Planted in the vicinity of Villages. Guna, Gwalior, Hosahangabad, Indore, Jabalpur, Mandla, Raisen, Seoni, Shivpuri, Sidhi.

Botany: Small trees, with 3-pinnate leaves and creamy-yellow or white flowers without red streaks on petals. Fruits slender, 3-gonous, 9-ribbed.

Flowering & Fruiting: February-June.

Medicinal Parts: Root, bark, leaf, seed.

Chemical constituents: Root bark contains Alkaloids, Moringine, Moringinine, Spirochin, Ptrigospermine.

Leaf & fruits contain Vitamin – A, Nicotinic acid, Ascorbic acid, Vitamin – B, fatty acid.

Flower & fruits contain Amino acids, glucose & sucrose, Citric acid, Malic acid, Succinic acid, Fumaric acid, Oxalic acid.

Medicinal Properties: Root – Acrid, anodyne, anthelmintic, anti – inflammatory, bitter, carminative, constpating, diuretic, dysentery, expectorant, stimulating, thermogenic.

Bark – Abortifacient, acrid, antifungal, bitter, thermogenic.

Leaf – Anodyne, anthelmintic, anti – inflammatory.

Seed – Acrid, anodyne, anti – inflammatory, antipyretic, bitter, Purgative.

Medicinal Uses: Root – Amenorrhoea, asthma, bronchitis, cough, diarrohoea, dysentery, dyspepsia, epilepsies, fever, hysteria, inflammation, otalgia, paralysis.

Bark – Ringworm

Leaf – Helminthiasis, inflammation, scurvy, tumours, wounds.

Seed – Fever, inflammation, neuralgia.

Ayurvedic preperations: *Shobhanjandi lep, Syamadi churna, Sudarshan churna, Sarvajwar-har louha, Shobhanjanadi kwath.*

280. *Morus alba* L. (Moraceae)

Vern.: Sehtut.

Distribution: Cultivated in orchards, gardens, parks and along roadsides. Chhatarpur, Hosahangabad, Sagar.

Botany: Deciduous tree. Leaves ovate to lanceolate, acute or accuminate, dentate serrate or lobed. Flowers monoecious. Male spikes elongated catkins. Female spikes short, ovoid. Fruits white or dark purple to red when ripe.

Flowering & Fruiting: January-May.

Medicinal Parts: Root, leaves, fruits.

Chemical constituents: Fruits contain Carotene (Vitamin – A), Thiamine, Nicotanic acid, Riboflavin, Ascorbic acid, Calcium, Phosphorus, Iron.

Medicinal Properties: Root – Anthelmintic, astringent, purgative.

Leaf – Antibacterial, diaphoretic, emollient.

Fruits – Aphrodisiac, cooling, diuretic, anthelmintic, laxative, sour, sweet.

Medicinal Uses: Leaf – Hoarsness.

Fruits – Burning sensation, diarrhoea, dyspepsia, lumbago, small pox.

Ayurvedic preperations: *Vanari gutika, Masabaladi pacana, Vajikarana ghirit, Brmhana gutika.*

281. *Mucuna pruriens* (L.) DC. (Fabaceae)

Syn.: *M. prurita* Hook.

Vern.: Kemuch.

Distribution: Climbing on Shrubs and trees on road sides and forest edges. Balaghat, Bhopal, Damoh, Dhar, Gwalior, Hosahangabad, Indore, Jabalpur, Mandla, Panna, Raisen, Seoni, Shahdol, Sidhi, W. Nimar.

Botany: Twining herbs, with 3-foliolate leaves and purple flowers in racemes. Pods densely bristly slightly curved at both ends.

Flowering & Fruiting: September-May.

Medicinal Parts: Root, leaf, seed.

Chemical constituents: Seed contains Carbohydrates, Steroids, Phenols, Glycosides, Tannin, Aluminium, Iron, Zinc, Calcium, Magnessium, Potessium.

Medicinal Properties: Root – Anthelmintic, aphrodisiac, bitter, diuretic, emollient, febrifuge, puragative, stimulant, sweet, thermogenic.

Leaf – Anthelmintic, aphrodisiac.

Seed – Anthelmintic, aphrodisiac..

Sour – Anthelmintic, aphrodisiac, astringent, laxative.

Medicinal Uses: Root – Amenorrhoea, constipation, dropsy, dysmenorrhoea, elephantiasis, fever, helminthiasis, ulcers.

Leaf – General debility, helminthiasis, inflammation, ulcers.

Sour – General deblility, gonorrhoea, sterility.

Ayurvedic preperations: *Vanari gutika, Masabaladi pacana, Vajikarana ghrit, Brmhana gutika.*

282. *Mukia madraspatana* (L.)Roem. (Cucurbitaceae)

Syn.: *Melothria madraspatana* (L.) Cogn.

Vern.: Bilari.

Distribution: Common throughout the State.

Botany: Annual climbers with wiry branches clothed with white hairs; Leaves orbicular, 3-5 lobed. Fruit globose, the size of pea, red when ripe; seeds 1-4 ellipsoids, turgid, white, and scabrous.

Flowering & Fruiting: July – December.

Medicinal Parts: Whole plant.

Chemical constituents: Root contain Columbin.

Medicinal Properties: Anodyne, bitter, carminative, expectorant, refrigerant, sweet.

Medicinal Uses: Asthma, burning sensation, constipation, cough, dyspepsia, flatulence, neuralgia, ulcers, vertigo.

283. *Murraya koenigii* (L.) Spr. (Rutaceae)

Syn.: *Bergera koenigi* L.

Vern.: Kathrim.

Distribution: Hoshangabad.

Botany: A small tree; Leaves imparipinnate, leaflets 11-25 alternate, obliquely ovate or rhomboid, irregularly crenate-dentate, pubescent beneath, sprinkled with black dots; flowers white, in corymbose cymes; fruits subglobose, black.

Flowering & Fruiting: December-June.

Medicinal Parts: Root, bark, leaf.

Chemical constituents: Plant contains Koenimb, Cadinine, Caryophyllene, Dipentene, Pinene, Phallandrene, Sabinene, Cavirol, Isosafrol, Terpineol, Lauric .acid, Palmitic acid, Girinimbine, Koenidine, Koenigine, Koeningicine, Koenimbidine, Koenine, Mahanine, Mukonicine.

Leaves contain Mahanimbidine, Mahanimbine, Curryanine, Curryangine & Murrayaline.

Medicinal Properties: Root – Acrid, anodyne, anthelmintic, anti – inflammatory, antiseptic, appetising, aromatic, astringent, bitter, carminative, constipating, cooling, depurative, febrifuge, stomachic.

Bark – Acrid, anodyne, anthelmintic, anti – inflammatory, antiseptic, appetising, aromatic, astringent, bitter, carminative, constipating, cooling, depurative, febrifuge, stomachic.

Leaf – Acrid, anodyne, anthelmintic, anti – inflammatory, antiseptic, appetising, aromatic, astringent, bitter, carminative, constipating, cooling, depurative, febrifuge, stomachic.

Medicinal Uses: Burning sensation, colic, diarrhoea, dysentery, dyspepsia, helmintic, inflammation, leprosy, leucoderma, pruritus, skin diseases, ulcer, vomit.

284. *Murraya paniculata* (L.) Jack. (Rutaceae)

Syn.: *M. exotica* L.

Vern.: Kamini.

Distribution: Bhopal, Damoh.

Botany: A small evergreen shrub or tree; leaves pinnately compound; flowers white, fragrant; fruits red berries.

Flowering & Fruiting: October-April.

Medicinal Parts: Leaves, Flowers, Fruits, Root.

Chemical constituents: Leaves contain – Phebalocin, Heptamethoxy flavon.

Fruits contain – Paniculatin, coumarin.

Flower contains – Scopoletin, glucose.

Roots contain – Alkaloids:Noracronycine & skimmiacine.

Medicinal Properties: Astringent, stimulant.

Medicinal Uses: Leaves – Diarrhoea, dropsy, dysentery.

Root bark – Cough, hysteria, rheumatism.

285. *Naravelia zeylanica*(L.) DC. (Ranunculaceae)

Distribution: In dense forests. On lower hill slopes. Hosahangabad.

Botany: Climbing tendriller shrubs, with fragrant Yellow flowers & aggregate of achenes fruits.

Flowering & Fruiting: November-February

Medicinal Parts: Whole plant.

Medicinal Properties: Anodyne, Anthelmintic, Anti-inflammatory Bitter, Depurative, Sweet, Vulnerary.

Medicinal Uses: Colic, Helminthiasis, Inflammation, Leprosy, Ulcers, Wounds.

286. *Nelumbo nucifera* Gaertn. (Nymphaeaceae)

Syn.: *Nelumbium speciosum* Willd.; *Nelumbium nelumbo* Druce.

Vern.: Kamal.

Distribution: In old tanks and Marshes. Damoh, Indore, Mandla, Sagar.

Botany: Perennial, rhizomatous aquatic herbs; leaves orbicular, peltate, 32-60 cm across; flowers white to pink, solitary on long axillary peduncles, sepals caducous; petals many seriate; carpels many, free, embeded in an obconical receptacle.

Flowering & Fruiting: April-January.

Medicinal Parts: Whole plant.

Chemical constituents: Rhizome contain Glucose, tannin, fat, resin, metarbin & an alkaloid – Nelumbine.

Medicinal Properties: Stem – Anthelmintic, astringent, cooling, diuratic.

Root – Bitter, cooling, diuratic, emollient.

Leaf – Bitter, cooling, diuratic, emollient.

Flowers – Refrigerant, sweet.

Fruits & Seeds – Aphrodisiac, astringent, bitter, cooling, diuratic, sweet.

Medicinal Uses: Stem – Helminthiasis, leprosy, skin diseases, vomit.

Root – Cough, diarrhoea, dyspepsia, leucoderma, small pox, spermatorrhoea.

Leaf – Burning sensation, fever, leprosy.

Flowers – Bronchitis, cholera, cough, diarrohea, fever.

Fruit & Seeds – Burning sensation, fever, leprosy, leucorrhoea, pruritus, vomit.

287. *Nerium indicum* Mill. (Apocynaceae)

Syn.: *N. odorum* Soland.; *N. oleander* L.

Vern.: Kaner.

Distribution: Ornamental, cultivated in gardens. Chhatarpur, Damoh, Dhar, Rewa, Satna, Shivpuri, Sidhi, W. Nimar.

Botany: Evergreen shrubs, with linear-elliptic leaves and white or red flowers in cymes.

Flowering & Fruiting: Throughout the year.

Medicinal Parts: Root, leaf.

Chemical constituents: Root bark contains Neriodorin, Nerioderin, Karabin, Glucoside (Odorin).

Leaf contains Oleandrin, Neriodin, Ursolic acid, Adynerin.

Flower contains Parafin, Benzyl alcohol, Compesterol, stigma sterol, fatty acids.

Medicinal Properties: Root/leaf – Acrid, anthelmintic, aphrodisiac, astringent, bitter, diuretic, febrifuge, stomachic, thermogenic.

Medicinal Uses: Root – Asthma, leprosy, pruritus, ulcers.

Leaf – Haemorrhoids, scabies.

Ayurvedic preperations:*Karvira yoga, Karviradi taila.*

288. *Nervilia aragoana* Gaud. (Orchidaceae)

Syn.: *Pogonia flabelliformis* Lindl.; *P. carinata* Wt.

Distribution: In Sal forests. Balaghat.

Botany: Terrestrial orchid with a single lotus like leaf, sepals & petals Pale-Green veined with Purple, lip White 3-lobed about the middle, veined with Purple, tinged with Yellow near the base, mid lobe ovate, acute, Crisped, Hairy.

Flowering & Fruiting: July-September.

Medicinal Parts: Whole plant.

Medicinal Properties: The plant is Bitter, Acrid, Cooling, Diuretic, Tonic.

Medicinal Uses: Agalactia, Asthma, Colic, Cough, Diarrhoea, Haemoptysis, Mental instability, Vomiting.

289. *Nicotiana tabacum* L. (Solanaceae)

Syn.: *N. macrophylla* Lehm.; *N. auriculata* Bert.

Vern.: Tambaku.

Distribution: Widely cultivated throughout the State and also as wild along the drains and on abandoned hill tops. Damoh, Hosahangabad, Mandla, Sidhi.

Botany: Glandular-pubescent herbs, with Pinkish flowers and ovate capsules.

Flowering & Fruiting: September-April.

Medicinal Parts: Leaves.

Chemical constituents: Leaves Nicotine alkaloid.

Medicinal Properties: Leaves – Abortifacient, acrid, anodyne, anthelmintic, anti-inflmmatory, bitter, carminative, emetic, laxative, narcotic, sedative, styptic, thermogenic.

Medicinal Uses: Leaves – Asthma, bronchitis, dyspepsia, flatlulence, haemorrhages, helminthiasis, inflammations, scabies, skin diseases, ulcers.

290. *Nyctanthes arbor-tristis* L. (Oleaceae)

Vern.: Harsinghar.

Distribution: Common in dry mixed deciduous forests and hill slopes. Balaghat, Chhatarpur, Damoh, Dhar, Indore, Jabalpur, Mandla, Narsimhapur, Panna, Rewa, Sagar, Satna, Seoni, Sidhi, W. Nimar.

Botany: Large shrubs, with 4-angled branches and ovate or ovate-lanceolate leaves. Flowers white to orange, fragrant. Capsule orbicular, compressed.

Flowering & Fruiting: September-January.

Medicinal Parts: Leaf, flower, seed.

Chemical constituents: Leaf & stem contains Beta – Sitosterol, Nyctanthic acid, Iridoid, Glycosides – Arbortristosides A, B, C, D & E Glucomannan, Glycerides of Lignocetric, Linocetric, Myristic acid, Oleic acid, palmitic acid, Stearic acid (In Seeds), As tragalin, Nictoflorin, Nyctanthoside, Arborsides, Ascorbic acid, Carotene, Glucose, Fructose, Manitol.

Medicinal Properties: Leaf – Acrid, anodyne, anthelmintic, antibacterial, anti – inflammatory, bitter, depurative, digestive, diuretic, expectorant, febrifuge, laxative, thermogenic.

Flower – Astringent, bitter, carminative stomachich.

Medicinal Uses: Leaf – Baldness, bronchitis, constipation, cough, dyspepsia, fever, helminthiasis, inflammation, pruritus, sciatica.

Seed – Scurvy.

291. *Nymphaea nauchali* Burm f. (Nymphaeaceae)

Syn.: *N.* lotus Hook. f. & Thoms non L.; *N. pubescens* Willd.; *N. rubra* Roxb. ex Salisb.

Vern.: Neelkamal.

Distribution: In tanks and marshes. Balaghat, Damoh, Gwalior, Indore, Jabalpur, Rewa, Sagar, Satna.

Botany: Rhizomatous perennial floating herbs; leaves large, 15-20 cm across; orbicular, regularly sinuately, toothed, densely tomentose beneath, flowers white to blue about 10 cm across.

Flowering & Fruiting: May-November.

Medicinal Parts: Rhizomes, flowers, seeds.

Medicinal Properties: Rhizome – Bitter, cooling, sweet, tonic.

Flower – Astringent, cardiotonic.

Seed – Aphrodisiac, constipation, cooling, stomachic, sweet.

Medicinal Uses: Rhizome – Diarrhoea, dysentery, dyspepsia, debility, gonorrhoea.

Seed – Diarrhoea, dispepsia.

292. *Ocimum basilicum* L. (Lamiaceae)

Syn.: *O. caryophyllantum* Roxb.; *O. pilosum* Willd.; *O. anisatum; Basilicum citratum.*

Vern.: Marwa.

Distribution: In wastelands, forest floor and nearby habitations. Balaghat, Betul, Chhatarpur, Gwalior, Indore, Jabalpur, Rewa, Satna, Seoni, Sidhi, Tikamgarh.

Botany: An erect branching herb. Stem and branches green or purplish. Leaves ovate-lanceolate, entire or toothed. Flowers whitish-pink or purplish, in whorls on simple or branched racemes. Corolla glabrous or variously pubiscent. Nutlets black, pitted.

Flowering & Fruiting: July-April.

Medicinal Parts: Whole plant.

Chemical constituents: Plant contains Ocimene, Methyl chavicol, Ceneol, Linoalool, Pinene – Camphor terpin hydrate, aromatic oil.

Medicinal Properties: acrid, anodyne, anthelmintic, antibacterial, antidiarrhoeal, antiemetic, anti-inflammatory, antipyretic, antispasmodic, appetiser, aromatic, bitter, carminative, depurative, digestive, diuretic, expectorant, insecticidal, pectoral, stimulating, stomachic, thermogenic.

Medicinal Uses: Affections, agalactia, amenorrhoea, asthma, bronchitis, cardiac debility, colic, cough, diarrohoea, dysentery, dyspepsia, fever flatulence, helminthiasis, inflammation, leucoderma, neuralgia, otalgia, pruritus, rheumatism, ringworm, skin diseases, spasmodic, vomit.

293. *Ocimum canum* Sims. (Lamiaceae)

Syn.: *O. americanum* Aut. non L.

Vern.: Jungli tulsi.

Distribution: In wastelands, forest floor, and nearby habitations. Chhatarpur, Indore, Mandla, Narsimhapur, Rewa, Satna, Seoni.

Botany: Herbs or undershrubs, with ovate-lanceolate or ovate-oblong leaves. Flowers white, whorled in long racemes. Calyx-tube hirsute-annulate within. Nutlets ovoid.

Flowering & Fruiting: August-March.

Medicinal Parts: Whole plant, leaves, seeds.

Chemical constituents: Plant contains Glycosides, alkaloids, tannins, savonin, volatile oil, vitamin – C, carotene.

Seed contains Green coloured oil, systosterol, fatty acid, palmitic acid, oleic acid, linoleic acid, linolenic acid, mucilage.

Medicinal Properties: Demulcent, expectorant, stimulant, stomachic.

Medicinal Uses: Antispasmodic, asthma, diabetes, itching, malaria, ringworm, spermatorrhoes, tuberculosis, urino – genital diseases.

294. *Ocimum gratissimum* L. (Lamiaceae)

Syn.: *O. frutescens*; *Citratum zeylanicum.*

Vern.: Ramtulasi.

Distribution: In Wastelands and along roadsides. Balaghat, Damoh, Dhar, Indore, Jabalpur, Seoni.

Botany: Tall, strongly aromatic shrubs; leaves ovate-lanceolate, tapering at base and decurrent on the petiole; flowers white, in verticillasters; nutlets 2 mm across.

Flowering & Fruiting: September– March.

Medicinal Parts: Whole plant, leaves, seeds.

Chemical constituents: Leaves contain Essential oil, steroids, ascorbic acid, glycosides, saponins, tannins.

Medicinal Properties: Alexiteric, appetiser, expectorant, heating, pungent, stomachic.

Medicinal Uses: Colic, cough, disuria, erysipelas, fever, gonorrhoea, inflammation, leucoderma, polyuria, seminal debility, skin diseases, strangury, vomiting.

295. *Ocimum kilimandscharicum Guerke.* (Lamiaceae)

Vern.: Kapur-tulsi.

Distribution: Indore.

Botany: Aromatic undeshrubs with pubescent quadrangular branchlets; leaves simple, opposite, oblong, narrow at the base, deeply serrate; flowers in 4-6 flowered whorls with a densely pubescent axis, corolla 2-3 times the length of the calyx; fruits ovoid-oblong nultlets.

Flowering & Fruiting: August-December.

Medicinal Parts: Leaves.

Chemical Constituents: Leaves yield an essential oil 6-13% of which 50-75% is camphor.

Medicinal Properties: Leaves – Acrid, Anti-bacterial, Antiviral, Appetising, Aromatic, Insecticidal, Ophthalmic, Thermogenic,

Medicinal Uses: Leaves – Anorexia, Bacterial & Viral infections, Bronchitis, Catarrh, Cough, Foul ulcers, Ophthalmopathy, Wounds.

296. *Ocimum sanctum* L. (Lamiaceae)

Syn.: *O. tenuiflorum* Sensu keng.; *O. hirsutum; O. tomentosum; O. virde.*

Vern.: Tulsi.

Distribution: Inwaste lands and under cultivation throughout the State. Chhatarpur, Damoh, Indore, Panna, Rewa, Sidhi, Tikamgarh.

Botany: A much branched, woody herb. Stems and branches clothed with soft spreading purplish hairs. Leaves opposite-decussate, ovate, elliptic-oblong, entire or crenate-serrate, hairy, minutely gland dotted. Flowers purplish-pink, whorld, in racemes, often forming panicles. Nutlets broad ellipsoid, smooth.

Flowering & Fruiting: Most part of the year.

Medicinal Parts: Whole plant.

Chemical constituents: Plant contains Essential oil Eugenol, Eugenol methyl ether, carvacrol, Caryophyllene, Ursolic acid, Apigenin, Luteolin, Ascorbic acid, Carotene, Alkaloid, Glycosides, Saponins & Tannins.

Medicinal Properties: Acrid, aromatic, bitter, diaphoretic, digestive, diuretic, expectorant, febrifuge, stomachic, vermifuge.

Medicinal Uses: Asthma, bronchitis, hiccough, leucoderma, lumbago, otalgia, ringworm, skin diseases, urino – genital disorders, vomit.

Ayurvedic preperations: *Tribhuvana kirti rasa, Caturbhuja rasa, Tulsi svarasa, Tulsi taila, Mukta pancamrta, Jvarasamharaka rasa.*

297. *Oldenlandia corymbosa* L. (Rubiaceae)

Vern.: Dhaman-paper.

Syn.: *Hedyotis corymbosa* (L.) Lam.

Distribution: Balaghat, Bhopal, Chhatarpur, Damoh, Gwalior, Hosahangabad, Jabalpur, Mandla, Narsimhapur, Panna, Ratlam, Sagar, Seoni, Shivpuri, Sidhi.

Botany: A small herbs; leaves subsessile, linear-lanceolate, acute, recurved and hairy margins, stipules membranous with few bristles; flowers white, small on solitary long peduncle; fruits globose capsule.

Flowering & Fruiting: September-April.

Medicinal Parts: Whole plant.

Chemical constituents: Whole plant – Alkaloids-Biflorine & Biflorone; Sodium, Potassium, Calcium, Chloride; caffeine, fumaric acid, oleanolic acid, ursolic acid, methyl ether.

Medicinal Properties: Anti-inflammatory, antipyretic, cooling, diuretic, expectorant, styptic.

Medicinal Uses: Bronchitis, fever, jaundice, liver disorders, malaria fever, measles.

298. *Operculina turpethum* (L.) Silva Manso. (Convolvulaceae)

Syn.: *Merremia turpethum* (L.) G.L. Shah & R.G. Bhat.; *Ipomoea turpenthum* R. Br.

Distribution: On bushes and hedges. Chhatarpur, Damoh, Mandla.

Botany: A large perennial twiner, with white flowers & capsule fruits.

Flowering & Fruiting: February-May.

Medicinal Parts: Roots.

Chemical constituents: Roots contain – Terpethin which is glycoside, Jalapine, Convolvulin, Soluble resin, Volatile oil, Lignin & Salts.

Medicinal Properties: Roots – Acrid, anthelmintic, antipyretic, bitter, carminative, expectorant, purgative, sweet, thermogenic.

Medicinal Uses: Roots – Bronchitis, colic, constipation, dropsy, erysipelas, helminthiasis, inflammation, jaundice, leucoderma, obesity, paralysis, pruritus, tumours.

299. *Opuntia elatior Mill.* (Cactaceae)

Syn.: *O. nigricans* Haw.

Vern.: *Naagphani.*

Distribution: In waste places, along roadsides & a hedges around cultivated fields.

Botany: Spreading bush, with flat, succulent, jointed, obovate branches having several straight spines on each cushion.

Flowering & Fruiting: March-Septeember.

Medicinal Parts: Fruits, Leaves, Latex.

Chemical Constituents: Flower contains Glycoside of Isorhamnetin and Quercertain and Glucoside narcissi. Isolation of Isoquercitrin, β-sitosterol, Opintiol. A Polysaccharide contains Galactose and Arabinose from Pods.

Medicinal Properties: Fruits – Refrigerant.

Latex – Purgative.

Medicinal Uses: Fruits – Gonorrhoea, Whooping Cough.

300. *Oroxylum indicum* (L.) Vent. (Bignoniaceae)

Syn.: *Bignonia indica* L.; *Calosanthes indica* Blume.

Vern.: Arlu.

Distribution: In mixed forests. Chhatarpur, Seoni.

Botany: Trees, with 2 to 3-pinnate leaves. Flowers in racemes; corolla fleshy; stamens 5. Capsules elongate, compressed.

Flowering & Fruiting: June – March.

Medicinal Parts: Root, leaf, fruit, seed.

Chemical constituents: Root bark contain Oroxylin – A bitter glycoside, acrid substance, pectin chrysie, baicalein oil.

Seed contain Tetuin, flavin, glucoside, fatty acid & Amino acids.

Bark contain Sodium salicilate.

Medicinal Properties: Root – Acrid, anodyne, anthelimintic, anti-Inflammatory, aphrodisiac, appetiser, astringent, bitter, carminative, diaphoretic, digestive, diuratic, expectorant, febrifuge, refrigerant, sweet.

Leaf – Anodyne, stomachic.

Fruits – Acrid, anthelmintic, stomachic, sweet.

Medicinal Uses: Root – Asthma, bronchitis, colic, cough, diarrhoea, dropsy, dysentery, dyspepsia, fever, flatulence, gout, hiccough, helmintic, inflammation, leucoderma, neuralgia, vomit.

Leaf – Flatulence, ulcers.

Fruits – Bronchitis, cardiac disorder, haemorrhoids, helminthiasis.

Ayurvedic preperations: *Syonaka siddha ghrit, Dashmula ghrit, Dashmula taila, Dashmularista, Chavyanprash, Shyonak putpak, Brhatpancamulyadi kwath.*

301. *Oryza sativa* L. (Poaceae)

Vern.: *Chawal.*

Distribution: Cultivated.

Botany: Tall, annual grass, panicles lax, at length drooping, spikelets persistent. Pale-brown at maturity.

Flowering & Fruiting: August-November.

Medicinal Parts: Roots, Grains.

Chemical Constituents: Grain contains Protein, Carbohydrate, Fibre; Minerals; Essential Amino acids, Free amino acids: Alamine, Tyrosine, Proline, Glutamic acid, Aspartic acid, Cystien, Vitamins.

Medicinal Properties: Grains – Acrid, Aphrodisiac, Diuretic, Sweet, Tonic.

Medicinal Uses: Grains – Biliousness.

302. *Ougeinia oojeinesis* (Roxb.) Hochr. (Fabaceae)

Syn.: *O. dalbergioides* Benth.

Vern.: *Tinsa.*

Distribution: In mixed forests. Balaghat, Bhopal, Chhindwara, Gwalior, Indore, Jabalpur, Mandla, Panna, Raisen, Sagar, Seoni, Shahdol.

Botany: A medium sized deciduous trees, with liliac or White flowers and flat pods.

Flowering & Fruiting: February-June.

Medicinal Parts: Bark.

Chemical Constituents: Bark contains – Triterpenoids as lupeol, Betulin, β-amyrin & Hydroxylupeol.

Medicinal Properties: Bark – Acrid, Astringent, Anthelmintic, Anti-inflammatory, Constipating, Cooling, Depurative, Febrifuge, Stimulant, Styptic, Sudorific.

Medicinal Uses: Bark – Anaemia, Burning sensation, Diarrhoea, Dysentery, Fever, Inflammation, Leprosy, Leucoderma, General debility, Gonorrhoea, Ulcers.

303. *Oxalis corniculata* L. (Oxalidaceae)

Vern.: Amrulsak.

Distribution: Balaghat, Hosahangabad, Jabalpur, Mandla, Narsimhapur, Panna, Sagar, Seoni, Sidhi.

Botany: Procumbent, apprised hairy herbs; leaves digitately 3-foliolate; leaflets subsessile cuneate or acute at base, appressed hairy; flowers yellow, 2-8 in umbellate clusters; capsule oblong, narrowed to the apex, pubescent; seeds transversely ribbed.

Flowering & Fruiting: January-December.

Medicinal Parts: *Chemical constituents:* Plant contains Malic acid, Tarteric acid, Citric acid, Potassium Oxalate.

Medicinal Properties: Anodyne, antibacterial, anti-inflammatory, antiseptic, astringent, carminative, constipating, diuratic, febrifuge, sour, thermogenic, vermifuge.

Medicinal Uses: Amenorrhoea, anemia, diarrhoea, dysentery, dysmenorrhoea, dyspepsia, fever, haemorrhages, scurvy, warts.

304. *Oxystelma esculentum* R. Br. (Asclepiadaceae)

Syn.: *O. secamone* (L.) karst.; *Periploca secamone* L.; *Periploca esculenta* L.f.

Vern.: *Dudhiyalata.*

Distribution: Climber on roadside shrubs. Balaghat, Hosahangabad.

Botany: A glabrous twiner. Leaves thin, narrowly lanceolate; flowers Whitish with Purple. Follicles glabrous, ovoid-lanceolate, pointed.

Flowering & Fruiting: September-December.

Medicinal Parts: Whole plant, Roots, Milky juice, Fruits.

Medicinal Properties: Whole plant – Anthelmintic, Aphrodisiac, Bitter, Diuretic, Dry, Hot, Indigestible, Lactagogue, Laxative, Pungent.

Fruits – Anthelmintic, Bitter, Expectorant, Tonic.

Medicinal Uses: Whole plant – Bronchitis, Leaucoderma, Sore throat, Ulcers.

Roots – Jaundice.

Fruits – Cough, Gleet, Gonorrhoea, Leucodema, Muscular pain.

305. *Pandanus fascicularis Lamk.* (Pandanaceae)

Syn.: *P. tectorius* Auct. Non Park.; *P. odoratissimus* L. f.

Vern.: Kevda.

Distribution: Common along streams and paddy fields.

Botany: Much-branched, small trees, with aerial roots. Leaves drooping spirally arranged, ensiform, with recurved marginal spines. Flowers white, sweet scented. Fruits woody, yellowish-brown.

Flowering & Fruiting: February-September.

Medicinal Parts: Root, leaf, flower.

Chemical constituents: Plant contains Physcion, Circilineol, Circilineol, n – tricontanol, β – stiosterol, stigmasterol, compesterol, Daucosterol, Palmitic acid, Stearic acid.

Medicinal Properties: Root – Acrid, anodyne, antiseptic, aphrodisiac, bitter, carminative, depurative, emollient, febrifuge, stomachic, sweet, thermogenic.

Leaf – Acrid, aphrodisiac, bitter, depurative.

Flower – Acrid, anodyne, aphrodisiac, bitter.

Medicinal Uses: Root – Abortion, colic, diabetes, dyspepsia, fever, flatulence, general debility, leprosy, otalgia, sterility, ulcers, wounds.

Leaf – Leprosy, leucoderma, scabies, small pox, syphilis, tumours.

Flower – Leucoderma, otalgia, pruritus.

Ayurvedic preperations: *Ark keora, Ketak panak, Ittre keora.*

306. *Panicum sumatrense* Roth. ex Roem. et Schult. (Poaceae)

Syn.: *Panicum miliare* Lam.

Distribution: Common throughout the State.

Botany: An annual grass.

Flowering & Fruiting: July-October.

Medicinal Parts: Grains.

Chemical Constituents: Grains – Albuminoids, Starch, Oil, Fibre.

Medicinal Properties: Grains – Alexeteric, Astringent, Constipating, Cooling, Digestive, Emollient, Sweet.

Medicinal Uses: Grains – Burning sensation, Diarrhoea, Dysentery, Dyspepsia, Gonorrhoea, Ulcers, Wounds.

307. *Paspalum scorbiculatum* L. (Poaceae)

Syn.: *P. commersonii* Lam.

Vern.: *Kodra.*

Distribution: Balaghat, Hosahangabad, Mandla, Shivpuri, Sidhi.

Botany: Erect or ascending, perennial grass, often rooting at lower nodes. Leaves tapering to a fine point. Racemes digitate, with 2-ranked spikelets. Caryopsis subglobose.

Flowering & Fruiting: August-September.

Medicinal Parts: Grains, Whole plant.

Chemical Constituents: *Grains contains* – Amino acids:Arginine, Histidine, Isoleucine, Leucine, Lysine, Methionine, Phenylalanine, Threonine, Tryptophan, Valine, N-lysine, Hentriacontanol, Hentriacontanone, Sitosterol, campesterol.

Medicinal Properties: Grains – Alexteric, Astringent, Bitter, Constipating, Cooling, Diuretic, Sedative, Styptic, Sweet, Tonic.

Medicinal Uses: Grains – Diarrhoea, Faltulence, Haemorrhages, Inflammation, Strangury, Ulcers.

308. *Pergularia daemia* (Forsk.) Chiov. (Asclepiadaceae)

Syn.: *P. extensa* (Jacq.) R. Br.; *Daemia extensa* (Jacq.) R. Br.

Vern.: Sadowani.

Distribution: Climber on roadside shrubs and trees. Balaghat, Bhopal, Chhatarpur, Damoh, E. Nimar, Indore, Mandla, Rewa, Sagar, Satna, Shahdol, Shivpuri, Tikamgarh.

Botany: Twining undershrubs, with cordate-ovate and acuminate leaves. Flowers yellowish-green, in drooping, corymbose cymes; corolla salver – shaped. Follicles echinate.

Flowering & Fruiting: May-July.

Medicinal Parts: Whole plant.

Medicinal Properties: The plant is acrid, anthelmintic, antipyretic, astringent, emetic, expectorant, laxative, thermogenic.

Medicinal Uses: Amenorrhoea, asthma, cough, dysmenorrhoea, fever, inflammation, leucoderma..

309. *Phaseolus mungo* (L.) Mant. (Fabaceae)

Syn.: *Vigna mungo* (L.) Hepper.

Vern.: *Urd.*

Distribution: Commonly cultivated. Chhindwara, Hoshangabad, Rewa, Sagar, Shivpuri.

Botany: Trailing, hispid herbs, with 3-foliolate leaves & yellow flowers. Pods linear, turgid.

Flowering & Fruiting: August-November.

Medicinal Parts: Whole plant, Roots.

Chemical Constituents: The plant contains Albuminoids, Starch and Oil.

Seed contains Saponin I, II & III. It contains Genistein, α-Hydroxygenistein, Dalbergioidin etc.

Medicinal Properties: Root – Narcotic.

Seed – Aphrodisiac, Appetiser, Diuretic, Laxative, Sweet, Tonic.

Medicinal Uses: Seeds – Asthma, Cough, Dysentery, Diarrhoea, Dyspepsia, Paralysis, Piles, Rheumatism.

310. *Phaseolus trilobatus* (L.) Schreb. (Fabaceae)

Syn.: *Vigna trilobata* (L.) Vend.; *Dolichos trilobatus* (L.) Mant.

Vern.: Mugam.

Distribution: In cultivated fields and roadsides. Balaghat, Bhopal, Damoh, Dhar, Gwalior, Hosahangabad, Indore, Jabalpur, Mandla, Narsimhapur, Panna, Raisen, Ratlam, Sagar, Satna, Seoni, Shivpuri, Sidhi, W. nimar.

Botany: A small perennial diffuse or trailing, prostrating, pubescent, 1-2 feet high herb; leaves trifoliate, each leaflet faintly 3-lobed and coriaceous; flowers in spikes on long peduncle, yellow pod-subcylindric.

Flowering & Fruiting: August-December.

Medicinal Parts: Whole plant.

Chemical constituents: Beans – Methionine, tryptophane, tyrosine strepogenin, uridine diphosphate, galacturonic acid.

Seeds – Lysine, valine, leucine, phenylalkanine.

Medicinal Properties: Anthelmintic, cooling, demulcent, febrifute, sedative, tonic.

Medicinal Uses: Antipyretic, cough, inflammations, Gout, tuberculosis.

311. *Phoenix dactylifera* L. (Arecaceae)

Vern.: Khajur.

Distribution: Planted. Bhopal.

Botany: A medium sized palm tree, the trunk covered by persistent petiole bases, the base surrounded by a dense mass of root suckers; leaves pinnately compound; flowers in long spadices, unisexual; fruits oblong berries.

Flowering & Fruiting: March-October.

Medicinal Parts: Leaves, Flowers, Fruits, Seeds.

Chemical constituents: Fruits Vitamin – A,B,D.

Pollen grains – Arabinose, galactose, xylose, rhamnose, uronic acid.

Medicinal Properties: Leaves – Aphrodisiac.

Flowers – Bitter, expectorant, febrifuge, purgative.

Fruit – Aphrodisiac, cooling, diuretic, sweet.

Toddy – Aphrodisiac, appetizing, digestive, intoxicating.

Medicinal Uses: Fruits – Bronchitis, burning sensation, cough, strangury.

Toddy – Bronchitis.

312. *Phoenix sylvestris Roxb.* (Arecaceae)

Vern.: Khajuri.

Distribution: Along roadsides and in villages. Bhopal.

Botany: Usually unbranched palms, with straight or crooked trunk & and large pinnate leaves; leaflets conduplicate, spine-tipped. Spathes woody, separating into boat-shaped valves. Drupes ellipsoid-oblong, orange-yellow.

Flowering & Fruiting: December-June.

Medicinal Parts: Fruits & Leaves.

Chemical constituents: Thiamine, riboflavin, nicotinic acid, vitamin-C, calcium, phosphorus, iron, sodium, potassium, copper, cobalt, nickel, magnessium, molybdenum.

Medicinal Properties: Fruit – Aphrodisiac, cardiotonic, constipating, cooling, restorative, sweet.

Medicinal Uses: Fruit – Abdominal complaints, fever, heart complaints, vomiting.

Root – Toothache.

313. *Phyla nodiflora* L. (Verbenaceae)

Vern.: Jalpippali.

Distribution: Common in moist wastelands. Chhatarpur, Damoh, Panna, Rewa, Satna, Shivpuri, Sidhi, Tikamgarh, W. Nimar.

Botany: A creeping, much branched perennial herb; stem 4-angled; leaves obovate, serrate-dentate margin, fleshy, hairy; flowers white or pale lilac, densely arranged into axillary heads; fruits globose-oblong, dry, splitting into 1-seeded glabrous pyrenes.

Flowering & Fruiting: Throughout the year.

Medicinal Parts: Whole plant.

Chemical constituents: Plant contain Fructose, glucose, lactose, xylose, linoleic acid, linolenic acid.

Leaf contain Nodiflorins – A & B, nodifloridins A & B, nepetin, lippiflorin – A, nodifloretin, glucosides of β-sterol & syigma-streol.

Medicinal Properties: The plant is acrid, anthelmintic, aphrodisiac, appetiser, astringent, constipating, cooling, digestive, diuratic, fever, stomachic.

Medicinal Uses: Asthma, bronchitis, colic, diarrhoea, dyspepsia, fever, flatulence, gonorrhoea, helmintic, ulcer.

314. *Phyllanthus emblica* L. (Euphorbiaceae)

Syn.: *Emblica officinalis* Gaertn.

Vern.: Amla.

Distribution: Common in mixed forests throughout the state, also planted for its edible fruits.

Botany: Large, deciduous trees, with distichous, linear-oblong leaves. Flowers greenish-yellow, in fascicles on leafless branches. Fruits globose, fleshy; seeds bony.

Flowering & Fruiting: March-December.

Medicinal Parts: Root, bark, leaf, fruit.

Chemical constituents: Fruits contain Protein, fats, fibres, Carbohydrates, Vitamin – C, Nicotinic acid, Tannins.

After drying fruits contain Gallic acid, Ellagic acid, Flavin & glucose.

Seed contains Linoleic acid, Linolenic acid, oleic acid.

Medicinal Properties: Root bark – Astringent.

Fruits – Acrid, alterant, anodyne, antipyretic, aphrodisiac, astringent, bitter, carminative, cooling, digestive, diuratic, laxative, sour, stomachic, sweet.

Medicinal Uses: Root bark – Diarrohoea, gonorrhoea, jaundice.

Leaf – Conjunctivitis, diarrohoea, dysentery, dyspepsia, inflammation.

Fruits – Anaemia, asthama, bronchitis, colic, cough, diabetes, diarrohoea, dysentery, dyspepsia, fever, flatulence, haemorrhages, inflammation, jaundice, leprosy, leucorrhoea, menorrhagia, peptie ulcer, skin diseases.

Ayurvedic preparations: *Chavyanprash, Brahmi rasayan, Dhatri louha, Triphala churna, Amalkyadi churna, Sudarshan churna, Sarvajwar louha, Agnitundi vati, Sanjivani vati, Phaltrikadi kwath, Punarnava mandur, Navayas louha, Rakta pittantak louha, Pippalayasav, Sringayadi churna, Amalaki rasayan, Karanjadi yoga.*

315. *Phyllanthus niruri* Auct. non. L. (Euphorbiaceae)

Syn.: *P. fraternus* Webster.; *P. amarus* Schum. & Thonn.

Vern.: Bhuiamla.

Distribution: Common as weed in gardens, fields and open sandy places near villages. Balaghat, Chhatarpur, Dhar, E. Nimar, Mandla, Panna, Shivpuri, Sidhi, Tikamgarh.

Botany: Erect, annual herbs; stem subterete or obscurely angled; leaves oblong; stipules lanceolate; male flowers 1-3 together; disc lobes 6, glandular; female flowers solitary; perianth segments oblong-spathulate; seeds brown.

Flowering & Fruiting: July – November.

Medicinal Parts: Root, fresh leaves, whole plant.

Chemical constituents: Leaves contain Bitter, Phyllanthin & Hypophyllanthin, Alkaloid, lignas, Vitamin – C, Quercetin, Astrogaln, Querscitrin, Rutin.

Root contains Glucoflavon.

Seed contains Oil which contains – Linoleic & Linolenic acid.

Medicinal Properties: Acrid, alexipharmic, bitter, cooling, diuretic, sour, sweet.

Medicinal Uses: Anaemia, anuria, asthama, biliousness, bronchitis, hiccough, leprosy, menorrhagia, urinary discharges.

Ayurvedic preperations: *Bhumyamalaki svarasa.*

316. ***Phyllanthus reticulatus Poir.*** (Euphorbiaceae)

Syn.: *Kirganelia reticulata* (Poir.) Ball.

Distribution: Common as hedges, in wastelands and deforested areas. Balaghat, Chhindwara, Damoh, Panna, Sidhi.

Botany: A large, diffuse or straggling, glabrous shrub. Branchlets drooping. Leaves oblong, elliptic or rotundate, glabrous. Male flowers Purplish, in fascicles. Female flowers solitary. Berries globose, dark Purple.

Flowering & Fruiting: September-June.

Medicinal Parts: Whole plant.

Chemical Constituents: Leaves & Bark contains – Tannic acid, Gum & a Crystalline Principle.

Medicinal Properties: Alterant, Astringent, Constipating, Cooling, Diuretic, Stomachic, Sweet.

Medicinal Uses: Burning sensation, Burns, Diarrhoea, Gastropathy, Obesity, Skin eruptions, Sores, Strangury.

317. *Physalis minima* L. (Solanaceae)

Vern.: Ban phutka.

Distribution: In waste lands. Balaghat, Betul, Chhatarpur, Damoh, E. Nimar, Hosahangabad, Mandla, Tikamgarh.

Botany: Erect herbs, with sinuate-dentate leaves. Flowers yellow, axillary, solitary; calyx 5-angled, enclosing yellow berries.

Flowering & Fruiting: August – January.

Medicinal Parts: Whole plant.

Chemical constituents: Plant contains Steroidal physalin A, B & C. Epaxyphysalin – B, Plamitic acid, stearic acid, Oleic acid, Linoleic acid, Hexadeconoid acid, Epoxy & Hydroxy fatty acid.

Medicinal Properties: Appetiser, bitter, cooling, diuretic, expectorant, luxative, sweet.

Medicinal Uses: Bronchitis, colic, cough, pruritus, ulcers.

318. *Piper betle* L. (Piperaceae)

Syn.: *Chavica betle* Miq.

Ver.: Paan.

Distribution: Often cultivated in many districts.

Botany: Perennial climbers, with cordate leaves and White flowers in Catkin inflorescence.

Chemical constituents: Leaves Nicotinic acid, ascorbic acid, carotin, essential amino acids.

Medicinal Properties: Plant – Acrid, anthelmintic, aphrodisiac, astringent, bitter, carminative, expectorant, febrifuge, laxative, stomachic, sweet.

Medicinal Uses: Plant – Asthma, bronchitis, cough, diarrhoea, dyspepsia, fever, leprosy, rheumatism.

319. *Piper longum* L. (Piperaceae)

Syn.: *Chavica roxburghii* Miq.

Vern.: Pipram.

Distribution: Rare. In moist shady places of mixed forests. Dhar, Sidhi.

Botany: Creeping undershrubs, rooting at nodes; flowers yellow, in elongate spikes; berries red when ripe.

Flowering & Fruiting: July-October.

Medicinal Parts: Roots, dried spikes.

Chemical constituents: Roots – Piperlongumine, piperlonguminine, sesamin, piperine, methyl-3, 4, 5-trimethoxycinnamate.

Medicinal Properties: Root – Anthlemintic, bitter, digestive, diuretic, expectorant, purgative, stomachic, thermogenic.

Dried spikes – Acrid, aphrodisiac, carminative, digestive, emollient, expectorant, febrifuge, laxative, stomachic.

Medicinal Uses: Root – Dyspepsia, gout, lumbago.

Dried spikes – Asthma, bronchits, colic, dyspepsia, epilepsy, fevers, flatulence, gonorrhoea. hiccough.

320. *Pisum sativum* L. (Fabaceae)

Syn.: *P. arvense* L.

Vern.: Matar.

Distribution: Commonly cultivated for seeds used as pulse. Bhopal, Chhatarpur, Damoh, Indore, Mandla, Sagar, Shahdol, Shivpuri, Sidhi.

Botany: A sub-erect or climbing annual. Leaflets ovate to oblong. Flowers usually white or Vari-coloured. On long peduncles. Pods broad seeds round, white.

Flowering & Fruiting: December-April.

Medicinal Parts: Seeds.

Medicinal Properties: Seeds – Acrid, cooling, depurative, emollient, laxative, sweet.

Medicinal Uses: Seeds – Bronchitis, burning sensation, cough, cuts, leprosy, measles, skin diseases, wounds.

321. *Plumbago zeylanica* L. (Plumbaginaceae)

Vern.: Chitrak.

Distribution: Along roadsides and in forest undergrowth; often planted in gardens. Balaghat, Damoh, Guna, Gwalior, Hosahangabad, Indore, Mandla, Rewa, Sagar, Seoni, Shahdol, Tikamgarh.

Botany: Erect or straggling undershrubs, with ovate-lanceolate leaves and white flowers in panicled racemes. Seed oblong, smooth.

Flowering & Fruiting: August-September.

Medicinal Parts: Root.

Chemical constituents: Plant contains Volatile oil, Chitranone, Alpha & Beta – Amyrin, Lupeol, Taraxasterol, Fructose, Glucose, Ivertase, Protease, Chloroplumbagin, Droserone, Elliptinone, Zeylanone, Zeylinone, Meritone, Catechol, Tannin, Amino acids, Plumbagic acid.

Root contains Plumgagin alkaloids.

Medicinal Properties: Root bark – Alterative, anthelmintic, astringent, Bitter, carminative, dry, hot, stomachic.

Medicinal Uses: Root – Anaemia, bronchitis, dysentery, inflammation, itching, leucoderma, liver diseases, piles.

Ayurvedic preperations: *Chitrakasiddha taila, Pancakolasava, Chitrakavaleha, Chittrakadi gutika, Chitrak haritaki, Chitrak ghrit, Chitrakadi churna, Devdaradi kwath, Amlyadi churna, Sudarshan churna, Sarvajwarhar louha, Agnitundi vati, Punarnava mandur, Navayai louha, Pipplyasav, Yograj guggul, Laghuvis-garvataila, Aryogavardhini vati, Chandraprabha vati.*

322. *Plumeria rubra* L. (Apocynaceae)

Syn.: *P. acuminata* Ait.; *P. acutifolia* Poir.

Distribution: Ornamental, planted in gardens and lawns for beautiful flowers.

Botany: An evergreen or partly deciduous trees; flowers cream coloured with Yellow centre.

Flowering & Fruiting: November-June.

Medicinal Parts: Root, Bark, Leaves, Latex.

Chemical constituents: Bark contains Fulvopluimerin plumieride, amyrin, Lupeol, bitter glucoside, essential oil, plumeric acid.

Medicinal Properties: *Root bark – Acrid, astringent, bitter, carminative, laxative, thermogenic.*

Medicinal Uses: Root bark – Leprosy, pruritus, ulcers.

323. *Pongamia pinnata* (L.) Pierre (Fabaceae)

Vern.: Karanj.

Distribution: Generally planted as an avenue tree. Balaghat, Betul, Bhopal, Damoh, Dhar, Hosahangabad, Indore, Jabalpur, Mandla, Panna, Rewa.

Botany: Trees, with imparipinnate leaves. Flowers pinkish-white. Pods obliquely oblong, woody, indehiscent, glabrous.

Flowering & Fruiting: March-February.

Medicinal Parts: Root, bark, leaf, flower, seed oil.

Chemical constituents: Seed contains Bitter fatty acid oil, Pongamol (Hongay oil) and protein.

Plant contains Alkaloids, tannins, sugar, resin, fatty oil (Karanjin).

Medicinal Properties: Root bark – Acrid, alexeteric, anthelmintic, bitter.

Leaf – Anthelmintic, digestive, laxative.

Seed – Anthelmintic, acrid, bitter, depurative.

Seed oil – Anthelmintic, depurative, styptic.

Medicinal Uses: Root – Gonorrhoea.

Bark – Beriberi, haemorrhoids, ulcers.

Leaf – Cough, diarrohoea, dyspepsia, flatulence, gonorrhoea, leprosy.

Seed – Anaemia, fevers, hydrocele, inflammation, lumbago, otalgia.

Seed oil – Leprosy, lumbago, ulcers, scabies.

Ayurvedic preperations: *Karanjadi churna, Karanjadi ghrit, Karanjadi taila, Karanjadi yog, Vidangadi lep.*

324. *Populus alba* L. ((Salicaceae)

Vern.: *Safed.*

Distribution: Bhopal, Jabalpur.

Botany: Erect deciduous trees, with pale Yellowish bark & leaves of wavy margins.

Flowering & Fruiting: March-July.

Medicinal parts: Bark, Leaf bud.

Chemical constituents: *Bark* – Flavonoid.

Bud – Salipopulin.

Leaves – Glucoside, Pupulin, Salicin.

Medicinal properties: *Bark* – Anti-periodic, Anti-rheumatic, Diuretic, Febrifuge, Tonic.

Medicinal uses: *Bark* – Blood purifier, Diarrhoea, Coryza, Debility, Gleet, Gonorrhoea, Intermittent fever, Neuralgia, Skin diseases, Strangury.

325. *Portulaca oleracea* L. (Portulacaceae)

Vern.: Kulfa.

Distribution: In open wastelands. Balaghat, Chhindwara, Gwalior, Indore, Jabalpur, Mandla, Panna, Rewa, Seoni, Shivpuri, Sidhi, Tikamgarh.

Botany: Diffuse herbs, with succulent, obovate-spathulate leaves and yellow flowers. Capsules obovoid.

Flowering & Fruiting: Throughout the year.

Medicinal Parts: Stem, leaf, seeds.

Chemical constituents: Plant contains Crude fibres, carbohydrates, calcium, magnesium, phasphorus, Iron, potassium, copper, sulphur, & chlorine.

Medicinal Properties: Stem & leaves – Alexeteric, alterant, antibacterial, bitter, diuretic, salty, sour, stomchic, thermogenic.

Seed – Antidysenteric, cooling, diuretic.

Medicinal Uses: Stem & leaves – Diabetes, dysenetery, dyspepsia, gonorrhoea, inflammation, jaundice, otalgia, scurvy, skin diseases, tumours, vomit.

Seed – Burns, dysentery.

326. *Portulaca quadrifida* L. (Portulacaceae)

Vern.: Loniya.

Distribution: A weed of waste places, In dry deciduous forests. Indore.

Botany: A small diffuse or prostrate annual. Stems filiform, glabrous, rooting at the nodes. Leaves sub-sessile, flat, and succulent. Flowers yellow, surrounded by a 4-5-leaved involucre and long silky hairs. Capsules conical.

Flowering & Fruiting: July-October.

Medicinal Parts: Leaves, stem.

Chemical constituents: Plant contains Starch, Saponin, Tannin, lignin, mucilage.

Medicinal Properties: Alterative, bitter, carminative, diuretic, laxative, sour, stomachic, vermifuge.

Medicinal Uses: Asthma, cough, eye diseases, fever, headache, inflammation, piles, skin diseases, ulcers, urinary discharges.

327. *Psidium guajava* L. (Myrtaceae)

Vern.: Amrud.

Distribution: Cultivated for its edible fruits, almost in all district.

Botany: A erect, woody shrub or small tree with reddish-brown wood and scaly, brownish bark.Leaves oblong or oblong-elliptic. Flowers on axillary peduncles of 1-3-flowered. Fruits berries.

Flowering & Fruiting: Almost throughout the year, mainly in rainy and cold season.

Medicinal Parts: Leaves, flower, fruits, root.

Chemical constituents: All parts contain Leucocyanidin, Querecetin, β – sitosterol.

Flower & leaves contain Ellagic acid, β – Caryophyllene, Guaijaverin, Umonene, Pinene, Niacin, Vitamin – C & B, Cadinene, β – Cophaene, Curcumene, Farnesene, Citric, Cinnamic, Malic, Oxalic, Quinic, Succinic, Tartaric acids, Riboflavin, Thiamine, Carotene, Tannins, sucrose, Glucose, Fructose, Rhamnose, Starch & Amino acids.

Medicinal Properties: Root – Anthelmintic, astringent, constipating.

Leave – Anodyne, antispasmodic, astringent, febrifuge.

Flower – Cooling, laxative

Fruits – Aphrodisiac, astringent, cooling, laxative, sour, sweet.

Medicinal Uses: Root – Diarrhoea, dysentery, haemorrhages, ulcers, vomit.

Leaves – Cholera, diarrhoea, epilepsy, ulcers, vomit, wounds.

Flower – Bronchitis, colic.

Fruits – Colic, diarrhoea, dyspepsia, dysentery, general debility.

328. *Psoralea corylifolia* L. (Fabaceae)

Vern.: Babchi.

Distribution: Bhopal, E. Nimar, Hosahangabad, Indore, Jabalpur, Raisen, Rewa, Sagar, Seoni, Shivpuri.

Botany: Annual herbs, with simple, suborbicular, gland-dotted leaves. Flowers bluish-purple, packed in upper part of axillary racemes. Pods oblong, 1-seeded.

Flowering & Fruiting: August – November.

Medicinal Parts: Seeds.

Chemical constituents: Seed contains Oil which contain Bakuchi Phenol (This contains Psoralane & Psoradin active Compound), Resin, Volatile oil.

Medicinal Properties: Seeds – Acrid, anthelmintic, aphrodisiac, bitter, diaphoretic, diuretic, laxative, stimulating, tomachic.

Medicinal Uses: Leprosy, leucoderma, ulcers.

Ayurvedic preperations: *Baucchi taila, Vidanga lep, Bakuci bija churna, Svitrahara vati, Khadirarista, Mahamanjisthadi kwath.*

329. *Pterocarpus marsupium* Roxb. (Fabaceae)

Vern.: Bijasal.

Distribution: In Sal and Mixed forests, also planted as avenue tree. Balaghat, Bhopal, Chhindwada, Damoh, Gwalior, Hosahangabad, Indore, Jabalpur, Mandla, Raisen, Rewa, Sagar, Seoni, Shahdol, Shivpuri, Sidhi.

Botany: Deciduous trees, with crooked stem and spreading branches. Leaves elliptic or ovate. Flowers pale-yellow, in fusco-paniculate racemes. Pods suborbicular, winged, glabrous.

Flowering & Fruiting: July-March.

Medicinal Parts: Heartwood, leaf, flower, gum.

Chemical constituents: Plant contains Protein, pentosan, mucilage, pterosupin, pseudobaptigenin, liquiritigenin, garbanzol, β – cudesmol, pterostilbene, marsupol, carpusin, propterol, marrsupinol, parsupin, oleanolic acid, tannins & kinotanic acid.

Medicinal Properties: Heartwood – Acrid, alterant, anodyne, anthelmintic, anti-inflammatory, astringent, bitter, constipating, cooling, depurative, rejuvenating, union-Promoter.

Flowers – Appetiser, bitter, cooling, febrifuge, sweet.

Gum – Anthelmintic, antipyretic, bitter, styptic.

Medicinal Uses: Heartwood – Asthma, bronchitis, diabetes, diarrhoea, dysentery, cough, elephantiasis, fractures, gout, haemorrhages, inflammation, leprosy, leucoderma, skin diseases.

Leaf – Boils, skin diseases, sores.

Flowers – Fevers.

Gum – Boils, diarrhoea, fever, helmintic, psoriasis, ulcers, wounds.

330. *Pueraria tuberosa* DC. (Fabaceae)

Vern.: Patal kumhra.

Distribution: On shrubs and small trees in mixed forests. Balaghat, Chhindwada, Hosahangabad, Indore, Jabalpur, Mandla, Shahdol.

Botany: A large spreading tuberous herbaceous twiner with very large tuberous roots; flowers blue or purplish blue in leafless racemes; fruits membranous, flat jointed pods clothed with silky brown hairs.

Flowering & Fruiting: February-April.

Medicinal Parts: Tuberous roots.

Chemical constituents: Hydroxytuberosone, anhydroxytuberosin, 3-O-methylandydrotuberosin, puerarostan, puerarone, daidzein, stigmasterol, puetuberosanol.

Medicinal Properties: Roots – Aphrodisiac, diuretic, emetic, emollient, expectorant, febrifuge, laxative, refrigerant, sweet.

Medicinal Uses: Root – Arthritis, burning sensation, constipation, cough, dyspepsia, fevers, leprosy, spermatorrhoea, strangury, tuberculosis.

331. *Punica granatum* L. (Punicaceae)

Vern.: Anar.

Distribution: Planted for its juicy seeds. Chhatarpur, Damoh, Sidhi.

Botany: A large deciduous shrub or small tree, with spiny or thorny branches and dark Grey bark.

Flowering & Fruiting: March-January.

Medicinal Parts: Roots, Bark, Flowers, Fruits, Seeds.

Chemical constituents: Bark – Alkaloid-Pelletierine.

Root – Alkaloids, pseudo-pelletierine, pelletierine, isopelletierine, methylpelletierine.

Flowers – Sitosterol, tannins-punicalagin & punicalin.

Fruits – Mannose, galactose, rhamnose, arabinose, glucose, galactouronic acid.

Medicinal Properties: Root & Stem bark – Anthelmintic, astringent, cooling.

Flowers – Styptic.

Fruits – Aphrodisiac, astringent, cooling, diuretic. laxative, sour, sweet.

Seed – Diuretic, stomachic.

Medicinal Uses: Root & Stem bark – Good for tapeworm and for strengthening the gums.

Flowers – Hydrocelc, ulcers, vomiting.

Fruits – Anaemia, bronchitis, otalgia.

Seed – Vomiting, scabies.

332. *Quisqualis indica* L. (Combretaceae)

Syn.: Q. densiflora Wall. ex. Miq.

Distribution: Commonly planted in gardens and houses for its attractive flowers. Bhopal, Jabalpur, Mandla, Shivpuri.

Botany: A large, scandent, shrubby and woody vine, with compact terminal spike inflorescence and capsule fruits.

Flowering & Fruiting: Throughout the year.

Medicinal Parts: Leaves, Seeds.

Chemical constituents: Leaves – Trigonelline, L-proline, L-asparagine, quisqualic acid.

Seed oil – Linolenic, oleic, palmitic, stearic, arachidic acids, sterol, fatty acids.

Medicinal Properties: Seeds – Anthelmintic, antidiarrhoeal, febrifuge.

Leaves – Carminative.

Medicinal Uses: Seeds – Diarrhoea, fever, helminthiasis, rickets, ringworm.

Leaves – Colic, constipation, helminthiasis.

333. *Randia dumetorum Poir.* (Rubiaceae)

Syn.: *Xeromorphis spinosa* Thunb. Keay.; *Catunaregum spinosa* (Thunb.) Tiru.

Vern.: Mainphal.

Distribution: Damoh, Hosahangabad, Satna, Shivpuri, Sidhi.

Botany: **–** A deciduous thorny shrub or small tree; Leaves elliptic to obovate, hairy on the veins beneath; flowers white, turning yellowish, solitary or fascicled in the axils; seeds embedded in pulp.

Flowering & Fruiting: May-January.

Medicinal Parts: Fruits.

Chemical constituents: 6 saponins-dumetoronins A, B, C, D, E & F – isolated, all of them contained oleanolic acid as glycone.

Medicinal Properties: Anodyne, anthlemintic, anti-inflammatory, carminative, expectorant, emetic, febrifuge, pungent, sudorific.

Medicinal Uses: Amenorrhoea, cold, cough, dysmenorrhoea, fever, flatulence, inflammations, piles, rheumatism, swelling, ulcers.

334. *Ranunculus sceleratus* L. (Ranunculaceae)

Vern.: *Jaldhania.*

Distribution: On moist river beds. Panna.

Botany: Annual herbs, with White flowers & achene fruits.

Flowering & Fruiting: December-March.

Medicinal Parts: Whole plant.

Chemical Constituents: Seed contain a Fatty oil.

Medicinal Properties: Whole plant – Acrid, Stimulant, Diuretic.

Medicinal Uses: Whole plant – Sciatica, Rheumatism, Dysuria, Asthma, Pneumonia.

Seeds – Tonic, Stomachic, Kidney troubles.

335. *Raphanus sativus* L. (Brassicaceae)

Syn.: R. caudatus L.

Vern.: Muli.

Distribution: Commonly cultivated as vegetable.

Botany: Annual herbs, with fusiform roots and white or lilac flowers.

Flowering & Fruiting: December-April.

Medicinal Parts: Roots, Leaves, Seeds.

Chemical constituents: Seeds – Glycoside, enzyme, methyl mercaptan, sulforaphene, essential oil.

Medicinal Properties: Roots – Acrid, anodyne, anti-inflammatory, appetizing, bitter, depurative, digestive, laxative, stomachic, thermogenic.

Leaves – Antibacterial, diuretic, laxative.

Seeds – Carminative, diuretic, expectorant, laxative.

Medicinal Uses: Root – Colic, diarrhoea, dyspepsia, flatulence, inflammations, leprosy, nausea, skin diseases.

Seeds – Amenorrhoea, cough, dyspepsia, flatulence, paralysis, strangury,

336. *Rhus parviflora* Roxb. (Anacardiaceae)

Vern.: Khatua.

Distribution: On higher hill slopes in mixed forests. Chhindwara, Hoshangabad.

Botany: A shrub, with Green flowers & drupe fruits.

Flowering & Fruiting: August-February.

Medicinal Parts: Fruits.

Medicinal Uses: Dental disorders, Conjuctivitis, Oedema.

337. *Ricinus communis* L. (Euphorbiaceae)

Vern.: Arandi.

Distribution: Usually cultivated for oil seeds, sometimes found as an escape in wastelands. Chhatarpur, Damoh, Dhar, Seoni, Sidhi, W. Nimar.

Botany: Large shrubs or small trees, with fistular stem and large, palmetly 5 to 10-lobed leaves. Flowers in pyramidal panicles. Capsules globose, spiny, of 3 – cocci; seeds caruncled.

Flowering & Fruiting: September-May.

Medicinal Parts: Root, Leaves, Flowers, Fruits, Seeds & Seed oil.

Chemical constituents: Root contains Resin, Tannin, starch.

Seeds contain 50% oil, protein, Ricinine, Ricin, Lipase enzyme, Fixed oil.

Fixed oil contains Glycerides, Ricinolesis, Isoricinoleic acid, stearic acid, Crystals of calcium oxalate, nitrate.

Medicinal Properties: Alterative, anti-inflammatory, aphrodisiac, carminative, febrifuge, purgative, stomachic.

Medicinal Uses: Fruits – Asthma, bronchitis, epilepsy, eye diseases, febrifuge, hydrocoel, jaundice, lumbago, piles, rheumatism, sciatica, skin diseases.

Leaves – Cough, worms infestation.

Ayurvedic preperations: *Erand pak, Erandmuladi kwath,Erandsaptak kwath, Rasanadi kwath, Rasna panchak kwath, Vishgarbh taila, Gandharva haritaki, Simhanada guggulu, Sadangu guggulu.*

338. *Rotula aquatica* Lour. (Boraginaceae)

Syn.: *Rhabdia lycioides* Sensu C.B. Clarke non Mart.

Vern.: *Dantratchu.*

Distribution: On rocky river beds and along streams. Balaghat, Damoh, Dhar, Indore, Jabalpur, Mandla, Rewa, Sagar, Seoni, Shahdol.

Botany: Small shrubs, with Pink or Reddish flowers and drupes fruits.

Flowering & Fruiting: October-February.

Medicinal Parts: Roots.

Medicinal Properties: Root – Astringent, Bitter, Cooling, Diuretic, Laxative.

Medicinal Uses: Root – Haemorrhoids, Diabetes & Venereal diseases, Renal and Vesical Calculi,

339. *Rumex crispus* L. (Poaceae)

Syn.: *Rumex dentatus* L.

Vern.: Ambavah.

Distribution: In moist places. Chhattarpur, Damoh, Hosahangabad, Indore, Jabalpur, Mandla, Panna, Rewa, Satna, Shahdol, Sidhi, Tikamgarh.

Botany: A perennial, erect glabrous herb.

Flowering & Fruiting: January-April.

Medicinal parts *Chemical constituents:* Emodin, Chrysophanic acid, Essential oil.

Medicinal properties: *Root* – Mildly Laxative, Astringent, Purgative.

Medicinal uses: *Seeds* – Diarrhoea.

340. *Rumex vesicarius* L. (Poaceae)

Distribution: Cultivated as vegetable, fodder crop & for its medicinal properties. Rewa.

Botany: An annual monoecious glabrous herb.

Flowering & Fruiting: **February-March.**

Medicinal parts: Whole plant.

Chemical constituents: Leaves contain large amounts of oxalates.

Medicinal properties: *Leaves* – Cooling, Appetizer, Diuretic, Astringent.

Seed – Cooling, Appetizer, Diuretic, Astringent.

Medicinal uses: *Leaves* – Snake bite.

Seed – Dysentery, Scorpion sting.

341. ***Rungia repens*** **Nees. (Acanthaceae)**

Vern.: Kharmor.

Distribution: Common along the roadside ditches. Balaghat, Damoh, Hosahangabad, Indore, Jabalpur, Mandla, Narsimhapur, Panna, Sagar, Satna, Seoni, Shivpuri, W. Nimar.

Botany: A spreading, procumbent herb, rooting from lower nodes. Leaves lanceolate to oblong-elliptic. Flowers Violet or Pinkish, in erect, terminal, Pretty spikes. Capsules hairy.

Flowering & Fruiting: November– January.

Medicinal Parts: Whole plant.

Chemical Constituents: Flowers, Luteolin-7-glycoside and Delphinidin, Besides Luteolin and another Flavone-Chrysoeriol.

Medicinal Properties: Plant – Diuretic, Vermifuge.

Medicinal Uses: Plant – Cough, Fever, Snake bite.

342. ***Saccharum arundinaceum Retz.*** **(Poaceae)**

Syn.: *Erianthus arundinaceus* (Retz.) Jesw. ex Heyne

Distribution: Throughout the State.

Botany: Found along River banks and Water logged areas. Stems glabrous, Smooth, Internodes, Very long; Leaves deep Green; Inflorescence large, Pink, White or Silvery; Spikelets in pair, one sessile & the other pedicelled.

Flowering & Fruiting: October November.

Medicinal Parts: Root, stem.

Medicinal Properties: Roots – Demulcent, Diuretic.

Stem – Aphorodisiac, Astringent, Depurative, Refrigerant, Sweet.

Medicinal Uses: Stem – Burning sensation, Dipsia, Erysipelas, Haemopathy, Uropathy.

343. *Saccharum munja Roxb.* (Poaceae)

Syn.: *Erianthus munja* Jesw.

Distribution: Throughout the State.

Botany: Perennial grass, pale violet or Green coloured-spikelete arranged in panicle inflorescence, fruits rectangular.

Flowering & Fruiting: October-November.

Medicinal Parts: Roots.

Medicinal Properties: Stem – Acrid, Aphrodisiac, Cooling, Sweet.

Medicinal Uses: Stem – Blood troubles, Burning sensation, Erysipelas, Eye diseases, Urinary complaints.

344. *Saccharum spontaneum* L. (Poaceae)

Distribution: Common. Balaghat, Bhopal, Gwalior, Hosahangabad, Indore, Mandla, Panna, Rewa, Sagar.

Botany: A very variable, tall, perennial, densely tufted grass. Leaves very long, narrow-linear, acuminate, coriaceous. Flowering panicles conical or lanceolate to oblong; peduncle softly silky just below the panicle. Spikelets paired, one pedicellate, the other sessile, 1-flowered, hermaphrodite.

Flowering & Fruiting: September-November.

Medicinal Parts: Roots.

Medicinal Properties: Roots – Aphrodisiac, Astringent, Diuretic, Laxative, Refrigerant, Sweet, Tonic.

Medicinal Uses: Roots – Burning sensation, Dysentery, Dysentery, Dyspepsia, General debility, Haemorrhoids, Menorrhagia, Renal & Vesical Calculi, Strangury.

345. *Salvadora persica* L. Var. *wightiana* [Planch ex Thw.] Verdc. (Salvadoraceae)

Syn.: *S. indica* Wight.

Distribution: Rare.

Botany: A small or medium sized, crooked or bent. Leaves variable in shape ovate, elliptic-ovate or ovate lanceolate, pale Green, glossy, fleshy. Flowers Greenish-Yellow, in axillary & terminal panicles. Drupes globose, fleshy, dark Red at Maturity.

Flowering & Fruiting: December-March.

Medicinal Parts: Bark, Leaves, Shoots, Fruits.

Chemical Constituents: Root – Alkaloid-Trimethylamine, β-Sitosterol, Glucotropaeolin, Monoclinic Sulphur.

Stem – β-Sitosterol, Octacosanol, Salvadoside, Salvadoraside, Siringin.

Leaves – Salvadoricine.

Medicinal Properties: Leaves – Anthelmintic, Antiscorbutic, Astringent, Bitter, Diuretic, Expectorant, Tonic.

Fruits – Acrid, Alexeteric, Aphrodisiac, Bitter, Digestive, Emollient, Purgative, Stomachic, Sweet, Thermogenic.

Medicinal Uses: Leaves – Asthma, Bronchitis, Constipation, Cough, Haemorrhoids, Scurvy, Strangury, Verminosis.

Fruits – Constipation, Gastropathy, Seminal weakness.

346. *Santalum album* L. (Santalaceae)

Vern.: Chandan.

Distribution: On hill slopes, usually cultivated. Dhar, Gwalior, Indore, Narsimhapur, Ratlam, Rewa.

Botany: Small, evergreen trees, with drooping branches. Leaves ovate-lanceolate. Flowers reddish-brown, in axillary and terminal cymes. Fruits globose, dark purple.

Flowering & Fruiting: September – December.

Medicinal Parts: Heart wood, oil.

Chemical constituents: Heart wood contains Resin, Tanic acid & Volatile oil which is thick, yellow & aromatic. It contains œ & β – santalol, Aldehyde, ketone, santenone, santalone.

Medicinal Properties: Wood – Alexeteric, antipyretic, aphrodisiac, bitter, cooling.

Medicinal Uses: Wood – Biliousness, bronchitis, burning sensation, heart diseases, small pox, vaginal discharges.

Ayurvedic preperations: *Chandanabalaksadi taila, Chandanadi vati, Chandanadi churna, Chandanasava, Sudarshan churna, Sadangodaka, Pippalasav.*

347. *Sapindus laurifolia Vahl.* (Sapindaceae)

Syn.: *S. trifoliata* Sensu Hiern non L.

Vern.: Reetha.

Distribution: In dry deciduous forests, also planted.

Flowering & Fruiting: November-April.

Botany: Medium sized deciduous trees, with White flowers & Drupe fruits.

Medicinal Parts: Root, Bark, Fruits.

Chemical Constituents: Acyclic Sequiterpene Oligoglycoside: Trifolioside II, Saponin A, Sapindoside B, Aglycones: Hederagenin, Oleanolic acid, Sapindic acid.

Seeds – Fatty oil: Glycerides Palmitin, Stearic, Eicosanin, Lignocerin, Olein, Cyanolipids Containing Oleic acids & Arachidic acid; Phenolic acids: Protocatechic acid, Cis – P-coumaric acid, P-hydroxy benzoic acid, Cinnamic acid.

Medicinal Properties: Roots & Bark – Demulcent, Expectorant.

Fruits – Abortifaceint, Acrid, Anthlemintic, Astringent, Bitter, Emetic, Expectorant, Thermogenic, Tonic.

Medicinal Uses: Roots – Epilepsy, Hysteria.

Fruits – Asthma, Cholera, Diarrhoea, Lumbago, Verminosis.

348. *Saraca asoca* (Roxb.) De Wilde. (Caesalpiniaceae)

Syn.: *S. indica* Auct. non L.

Vern.: Ashok.

Distribution: Planted as an ornamental. Bhopal, Jabalpur, Sagar.

Botany: A small evergreen tree. Leaves glabrous cariaceous; leaflets oblong or oblong-lanceolate, slightly oblique. Flowers fragrant. Calyx petaloid, tubular. Anthers purple. Ovary pubescent. Pods black, linear-oblong, veined.

Flowering & Fruiting: December-July.

Medicinal Parts: Bark, flower, seed.

Chemical constituents: Bark contains Glycosides, Saponins, Tannins, Essential oil, Haematoxyline, Ketosterol, Kaempferol, Calcium, Iron, Fatty acids, Gallic acid.

Medicinal Properties: Anthelmintic, anti-inflammatory, astringent, blood purifier, cooling, demulcent, febrifuge, refrigerant, spasmodic, stimulant.

Medicinal Uses: Anti-bacterial, anti-cancer, anti-fungal, anti-tumour, cardiac stimulant, disuria, leucoderma, menorrhagia, piles, ulcers.

Ayurvedic preperations: *Ashokarishta and Ashokaghrit, Ashoka vati, Ashokadi kwath.*

349. *Sarcostemma acidum* (Voigt) Voigt (Asclepiadaceae)

Syn.: *S. brevistigma* Wight & Arn.

Vern.: Somlata.

Distribution: Rarc. Creeping on steep rocky slopes. Hosahangabad.

Botany: Perennial leafless, jointed shrubs, with milky latex and fragrant White flowers.

Flowering & Fruiting: July – January.

Medicinal Parts: Whole plant.

Chemical Constituents: Lupeol acetate, □-amyrin acetate, β-amyrin-acetate, β-sitosterol, Malic acid, Succinic acid, Sucrose, Tannins, Sarcobiose, Brevobiose, Tigmobiose.

Medicinal Properties: Whole plant – Acrid, Alterant, Bitter, Cooling, Emetic, Narcotic, Rejuvenating.

Medicinal Uses: Whole plant – Dipsia, General debility, Hydrophobia, Viral infection.

350. *Schleichera oleosa* (Lour.) Oken. (Sapindaceae)

Syn.: *S. trijuga* Willd.

Vern.: Kusum.

Distribution: Common in dry deciduous forests, often along river banks. Bhopal, Hosahangabad, Mandla, Satna, Seoni, Shahdol.

Botany: A handsome tree. Leaves paripinnate. Flowers Yellowish-Green. Fruits pointed, Smooth or echinate.

Flowering & Fruiting: March-August.

Medicinal Parts: Bark, Oil.

Chemical Constituents: Seed – Phenols, Polyene Pigments.

Seed oil – Cyanolipids, Oleic acid, Gadoleic acid, Stearic acid, Arachidic acid, Behenic acid, Palmitic acid, Palmitoleic acid.

Leaves – Stigmasterol, β-sitosterol, Campesterol, Cholesterol, Fructose, Glucose, Tartaric acid, Oxalic acid.

Medicinal Properties: Bark – Antipyretic, Astringent.

Seed oil – Anthelmintic, Bitter, Purgative, Sour, Stomachic, Sweet, Tonic.

Medicinal Uses: Bark – Inflammation, Malaria, Pruritus, Ulcers.

Seed oil – Burns, Cephalalgia, Ulcers.

351. *Schrebera swietenioides Roxb.* (Oleaceae)

Vern.: Banpalash.

Distribution: Common in mixed forests. Hosahangabad, Indore, Jabalpur, Mandla, Morena, Rewa, Sagar, Seoni, Sidhi.

Botany: Moderate sized deciduous tree, with Yellowish-Brown flowers & capsule fruits.

Flowering & Fruiting: March-December.

Medicinal Parts: Root, Bark, Leaves, Fruits.

Chemical Constituents: Leaves – Acetic acid, Formic acid, Methyl alcohol, β-sitosterol.

Medicinal Properties: Root / Bark / Leaves – Bitter, Acrid, Appetising, Digestive, Thermogenic, Stomachic, Depurative, Constipating, Urinary astringent, Anthelmintic.

Fruits – Digestive, Purgative, Stomachic.

Medicinal Uses: Root / Bark / Leaves – Anaemia, Boils, Burns.

Colic, Diarrhoea, Dyspepsia, Flatulence, Leprosy, Skin diseases.

Fruits – Anaemia, Anorexia, Colic, Diabetes, Flatulence.

352. *Scirpus grossus* L.f. (Cyperaceae)

Syn.: *S. kysoor* Roxb.

Distribution: Sagar.

Botany: Shrubs, with Brown flowers & drupe fruits.

Flowering & Fruiting: September-December.

Medicinal parts: Rhizomes & Flowers.

Chemical constituents: Steroids & Proteins, Carbohydrates.

Medicinal properties: *Root* – Astringent. Nutritive, Diuretic.

Medicinal uses: *Root* – Diarrhoea, Vomiting, Eye diseases, Cholera, Seminal debility, Miscarriage, Cholera.

Flowers – Jaundice.

353. Securinega leucopyrus (Willd.) Muell-Arg. (Euphorbiaceae)

Syn.: *Fluggea leucopyrus* Willd.

Vern.: Hartho.

Distribution: Rare, on hill slopes. Hosahangabad.

Botany: A large, woody, much branched, diffuse shrub. Branches White; branchlets spine-like, Purplish when young. Leaves less than 2.5 X 1.5 cm., obovate, elliptic or obcordate, glaucous. Male flowers Greenish-Yellow, slender, pedicelled Berries globose, White.

Flowering & Fruiting: February-November.

Medicinal Parts: Whole plant.

Chemical Constituents: Leaves contain – Triterpenes, Diterpenes, Steroids.

Stem bark – Tannins, Friedelin.

Medicinal Properties: Whole plant – Aphrodisiac, Cooling, Diuretic, Sweet, Tonic.

Medicinal Uses: Whole plant – Burning sensation, General debility, Seminal weakness, Strangury.

354. *Semecarpus anacardium* L. f. (Anacardiaceae)

Vern.: *Bhelwa.*

Distribution: Common in dry deciduous & Sal forests. Balaghat, Bhopal, Hosahangabad, Jabalpur, Mandla, Rewa, Seoni, Shahdol.

Botany: A small trees, with unisexual flowers and drupe fruits.

Flowering & Fruiting: May-March.

Medicinal Parts: Fruits.

Chemical Constituents: *Fruits contain* – Anacardiac acid, Cardol, Catechol, Anacardol, Fixed oil, Somearpol, Bhilawanol.

Medicinal Properties: Fruits – Acrid, Anthelmintic, Anti-Inflammatory, Aphrodisiac, Astringent, Bitter, Depurative, Digestive,

Emollient, Expectorant, Febrifuge, Purgative, Sudorific, Sweet, Thermogenic, Tonic.

Medicinal Uses: Fruits – Asthma, Caridac diseases, Colic, Constipation, Cough, Diabetes, Dyspepsia, Fever, Flatulence, General debility, Haemorrhoids, Helminthiasis, Hepatopathy, Sciatica, Scrofula, Tumours, Ulcers.

355. *Senna absus* (L.) Roxb. (Caesalpiniaceae)

Syn.: *Cassia absus* L.

Vern.: Chaksu.

Distribution: In waste lands. Balaghat, Bhopal, Damoh, E. Nimar, Indore, Mandla, Panna, Raisen, Sagar, Satna, Seoni, Shahdol, Shivpuri, Tikamgarh.

Botany: Annual herbs, clothed with glandular hairs; rachis grooved; leaflets obliquely elliptic to obovate; flowers pink or red, in terminal or leaf opposed, few-flowered racemes; pods hairy, flat.

Flowering & Fruiting: August –November.

Medicinal Parts: Leaves, seeds.

Chemical constituents: Seed contains Alkaloids, sugars, fats, cannins, mucilage, major protein, mangenese.

Medicinal Properties: Leaf – Acrid, astringent, bitter, constipating, expectorant, haematinic.

Seed – Acrid, astringent, bitter, constipating, diuretic, haematinic, vulnerary.

Medicinal Uses: Leaf – Anaemia, asthma, cough, diarrhoea, hiccough, nasitis, tumours.

Seed – Diarrhoea, dysentery, haemorrhages, obesity, poisonous bites, ulcers, wounds.

356. *Senna auriculata* (L.) Roxb. (Caesalpiniaceae)

Syn.: *Cassia auriculata* Roxb.

Vern.: Amoli, Chhoti.

Distribution: Gwalior, Hosahangabad, Indore, Shivpuri, W. Nimar.

Botany: Finely pubescent shrubs, leaflets 5 – 12 pairs, oblong-obovate, obtuse or emarginate, apiculate; flowers in corymbose racemes, yellow; pods oblong, mucronate, deeply impressed and septate between the seeds.

Flowering & Fruiting: July-December.

Medicinal Parts: Root, bark, leaves, flowers, seeds.

Chemical constituents: Plant contains Tannins.

White bark contains More tannic acid.

Medicinal Properties: Root – Astringent, cooling, depurative.

Leaf – Anthelmintic, depurative.

Seed – Anthelmintic, aphrodisiac, astringent, constipating, cooling, depurative, sour, stomachic.

Medicinal Uses: Root – Asthma, leprosy, skin disease, tumours.

Leaf – Leprosy, skin diseases, ulcers.

Flower – Diabetes.

Seed – Abdominal disorders, diabetes, diarrhoea, dysentery, leprosy, ophthalmia, skin diseases, swelling.

357. *Senna occidentalis* (L.) Link. (Caesalpiniaceae)

Syn.: *Cassia occidentalis* L.

Vern.: Ban-chironta.

Distribution: In Wastelands. Bhopal, E. Nimar, Hosahangabad, Indore, Mandla, Panna, Raisen, Ratlam, Rewa, Sagar, Satna, Sidhi, Shivpuri, W. Nimar.

Botany: Diffuse undershrubs, with 3-6 pairs of leaflets and a gland on the petiole. Flowers yellow, in corymbose panicles. Pods linear-oblong.

Flowering & Fruiting: July-April.

Medicinal Parts: Whole plant.

Chemical constituents: Plant contains Emodin, Tannic acid, fatty oils, Mucilage, aminoacids, Chrysophanol, Sitosterol, Xanthon, Cassiodin, Nitrogen bases, Betain, Choline, Stachydrine, and Trigonelline.

Seed contains Galactomanon.

Medicinal Properties: Root – Acrid, anti – inflammatory, bitter, digestive, diuretic, stomachic.

Leaf – Acrid, aphrodisiac, bitter, depurative, expectorant.

Seed – Acrid, bitter, depurative, diuretic, expectorant, febrifuge, purgative, stomachic.

Medicinal Uses: Root – Colic, diabetes, dyspepsia, elephantiasis, epilepsy, flatulence, inflammation, ringworm, scorpion sting.

Leaf – Asthma, bronchitis, cough, fever, hiccough, hydrophobia, leprosy, pruritus, ulcers, wounds.

Seed – Bronchitis, constipation, cough, dyspepsia, erysipelas, fever, flatulence, hiccough, leprosy, stranguray, ulcers.

358 Senna tora (L.) Roxb. (Caesalpiniaceae)

Syn.: *C. tora* L.; *C. obtusifolia* L.; *C. toroides* Roxb.; *C. foetida* Pers.

Vern.: Chakunda.

Distribution: Common in Wastelands. Balaghat, Chhindwara, Damoh, E. Nimar, Hosahangabad, Indore, Jabalpur, Mandla, Panna, Raisen, Sagar, Shivpuri, Tikamgarh.

Botany: Suffruticose herbs. With 3 pairs of leaflets and glands between them. Flowers yellow or reddish tinged, geminate; all 7 anthers rounded at apex. Pods linear, turgid.

Flowering & Fruiting: August-April.

Medicinal Parts: Leaves, seeds.

Chemical constituents: Seed contains Glucosides, (Emodin) proteins, steroids, Amino acids, Reducing sugars, Triterpenes, Tannins, oleic acid, Linotic acid, Palmitic acid, Lignocenic acid, Rubrofusarin, Fixed oils, sodium, Potassium & Iron.

Medicinal Properties: Leaves & seeds – Acrid, anthelmintic, antiperiodic, depurative, expectorant, laxative, ophthalmic.

Medicinal Uses: Leaves & seeds – Bronchitis, cardiac disorders, colic, constipation, cough, dyspepsia, flatulence, intermittent fever, leprosy, pruritus, ringworm, skin diseases.

359. *Sesbania grandiflora* (L.) Poir.(Fabaceae)

Syn.: *Agati grandiflora* Desv.; *Robinia grandiflora* L.

Vern.: Agastya.

Distribution: Usually planted in gardens. Bhopal, Chhatarpur, Indore, Sagar, Seoni.

Botany: Tall slender tree; Leaves pinnately compound, leaflets 16-30 pairs, linear-oblong. Flowers white.

Flowering & Fruiting: September-March.

Medicinal Parts: Root, Bark, Leaves, Flowers, Fruits.

Medicinal Properties: Bark – Astringent.

Leaves – Alexeteric, anthelmintic, bitter, indigestible.

Flowers – Appetizer, cooling.

Fruits – Acrid, alexeteric, bitter, laxative.

Medicinal Uses: Leaves – Epilepsy, gout, itching, leprosy, night blindness.

Flowers – Biliousness, bronchitis, gout, night blindness.

Fruits – Anaemia, bronchitis, fevers, pains, tumours.

Ayurvedic preperations: *Agastya ghrit, Agastyavaleha, Agastya prasa, Agasti puspa ghrit, Agastyadi lep.*

360. *Sesamum indicum* L. (Pedaliaceae)

Vern.: *Til.*

Distribution: Often cultivated for oil yielding seeds. Balaghat, Chhindwara, Damoh, Dhar, Indore, Mandla, Panna, Rewa, Satna, Seoni, Sidhi.

Botany: An erect, hairy robust or tall herb. Lower leaves often compound, lobed or deeply devided, broad ovate; upper ones undevided, linear-oblong. Flowers Pink or rosy, bent, solitary; capusules erect oblong, valvular.

Flowering & Fruiting: September-October

Medicinal Part: Roots, Leaves, Seeds, Oil.

Chemical Constituents: Leaves – A flavonoid as Pedailin & Pinoresinol.

Seeds **–** Sesamol, β-tocopherol, a lignan as Sesamin; a Nitrolactone as Sesamolin, Thiamine, Niacin, Globulin, Sesamolinol, Sesaminol, Pinoresinol, Sesamolinol.

Medicinal Properties: Roots / Leaves – Emollient.

Seeds – Acrid, Aphordisiac, Astringent, Bitter, Digestive, Emollient, Laxative, Sweet, Thermogenic, Tonic.

Oil – Anthelmintic, Astringent, Bitter, Constipating Digestive, Emollient, Sweet, Thermogenic.

Medicinal Uses: Leaves – Cholera, Dermatopathy, Dysentery, Nephropathy, Ophthalmopathy.

Seeds – Burns, Dysentery, Haemorrhoids, Migrain, Obesity, Ulcers.

Oil – Burning sensation, Cephalalgia, Gonorrhoea, Obesity, Otalgia.

361. *Setaria italica* (L.) P. Beauv. (Poaceae)

Distribution: Cultivated throughout the state.

Botany: Herbaceous annual plants, with spikelets inflorescence & various coloured oblong or ellipsoid grains.

Flowering & Fruiting: July-October.

Medicinal Parts: Grains.

Chemical Constituents: Grains – Folic acid, Necotinic acid, Riboflavin, Thiamine, Albumin, Globulin, Glutenin, Prolamine, □-setarin, β-setarin, Alanine, Lutamic acid, Histidine, Isoleucine, Leucine, Lysine, Methionine, Proline, Phenylalanine, Threonine, Tryptophane, Valine, Phytic acid, Arachidic acid, Behemic acid, Linoleic acid, Oleic acid, β-alanine, Betaline.

Medicinal Properties: Grains – Acrid, Aphrodisiac, Astringent, Cooling, Diuretic, Laxative, Sweet.

Medicinal Uses: Grains – Burning sensation, Fractures, Rheumatism.

362. *Shorea robusta* Gaertn. f. (Depterocarpaceae)

Vern.: Sal, Sarai.

Distribution: In well drained soil on lower hill slopes. Balaghat, Hosahangabad, Mandla, Rewa.

Botany: A large deciduous trees; flowers cream or Pale yellow, in racemes.

Flowering & Fruiting: March-June.

Medicinal Parts: Bark, leaves, resin & fruits.

Chemical constituents: Seed – Fatty oil.

Wood – Oleoresin.

Medicinal Properties: Bark & Leaves – Acrid, alexeteric, anodyne, anthelmintic, astringent, cooling, diuretic.

Resin – Acrid, anodyne, aphrodisiac, astringent, carminative, cooling, detergent, diuretic, expectorant, stomachic.

Fruits – Aphrodisiac, astringent, cooling, sweet.

Medicinal Uses: Bark, Leaves, Seed – Anaemia, asthma, bleeding piles, cough, diabetes, diarrhoea, dysentery, ear diseases, fracture, gonorrhoea, leprosy, sprain, wounds.

Fruit – Burning sensation, seminal weakness, tubercular ulcers.

Resin – Burns, diarrhoea, dysentery, fever, gonorrhoea, pruritus, ulcers, wounds.

363. *Sida acuta* Burm. (Malvaceae)

Syn.: *S. carpinifolia* L. f.

Vern.: Mahabala.

Distribution: In wastelands, fallow fields, and along roadsides and forests. Balaghat, Damoh, Gwalior, Hosahangabad, Mandla, Raisen, Rewa, Sagar, Shivpuri, Sidhi.

Botany: Undershrubs; flowers Yellow, solitary axillary.

Flowering & Fruiting: August-January.

Medicinal parts: Roots

Chemical constituents: *Root contains* – Alkaloids as β-Phenethylamine, Ephedrine, Vasicinol, Vasicinone, Vasicine, Choline, Hypaphorine, Betaine, Cryptolepine, □ -Amyrin.

Medicinal properties: *Root* – Astringent, Aphrodisiac, Demulcent, Febrifuge, Stomachic, Tonic.

Medicinal uses: *Root* – Diabetes, Dysuria, Paralysis, Leucorrhoea, Nervous disorders, Sciatica, Tuberculosis, Urinary disorders.

364. *Sida cordata* Borsm. (Malvaceae)

Syn.: *S. humilis* cav.; *S. veronicifolia* Lam.

Distribution: In Wastelands and forest margins. Balaghat, Bhopal, Damoh, Hosahangabad, Indore, Jabalpur, Mandla, Raisen, Rewa, Sagar, Satna, Seoni, Shivpuri, Sidhi.

Botany: Trailing hispid herbs, with solitary axillary Yellow flowers & schizocarpic fruits.

Flowering & Fruiting: *Throughout the year.*

Medicinal parts: Whole plant.

Chemical constituents: β-pheneethylamines; Quinazoline, Carboxylated tryptamine, Linoleic acid, Malavalic acid, Sterculic acid, Gossypol, Ephedrine, Vasicinol, Vasicinone, Vasicine, Choline, Betaine, Cholesterol, Stigmasterol.

Medicinal properties: *Root* – Astringent, Acrid, Bitter, Sour, Sweet, Thermogenic.

Leaves – Antidiarrhoeal.

Flowers / Ripe fruits – Refrigerant.

Medicinal uses: *Root* – Fever, Arthritis.

Root bark – Gonorrhoea, Leucorrhoea.

Leaves – Diarrhoea.

Flowers / Ripe fruits – Burning sensation.

365. *Sida cordifolia* L. (Malvaceae)

Vern.: Bala.

Distribution: In wastelands, forest margins and roadsides. Balaghat, Bhopal, Hosahangabad, Mandla, Seoni, Shivpuri, Sidhi.

Botany: Undershrubs, with ovate-cordate leaves and axillary yellow flowers. Mericarps 10, 2-awned.

Flowering & Fruiting: August– November.

Medicinal Parts: Root, leaves, seeds, whole plant.

Chemical constituents: Root contains Beta-sitosterol, alkaloids – visicinonoe, vasicine, vasicinol, choline, & betain.

Seeds contain Ephedrine, fatty oils, steroids, resin, mucin, sterculic acid, malvalic acid, linoleic acid, potassium nitrate.

Medicinal Properties: Plant – Aphrodisiac, astringent, bitter, emollicent, sweet, tonic.

Fruits – Acrid, aphrodisiac, astringent, cooling, digestive, sweet.

Medicinal Uses: Plant – Cough.

Bark – Urinary trouble & discharges.

Fruits – Blood diseases, diseases of throat, piles.

Ayurvedic preperations: *Bala taila, Satapakiksirabala, Bala churna/Svarasa, Baladya ghrit, Baladyarista, Baladi Kwath, Chandanabalalaksadi taila, Baladya ghrit, Baladyarista, Sudarsana churna, Kukuvadi churna.*

366. *Sida rhombifolia* L. (Malvaceae)

Vern.: Atibala.

Distribution: In wastelands, rock-crevices, forests and along streams. Balaghat, Bhopal, Hosahangabad, Mandla, Seoni, Shivpuri, Sidhi.

Botany: Much-branched herbs, with obovate – rounded or rhomboid leaves and axillary, solitary, yellow flowers. Mericarps 1-awned or awnless.

Flowering & Fruiting: October – December.

Medicinal Parts: Whole plant, root, leaves.

Chemical constituents: Plant contains Ephedrine, cellulose, lignin, quanozolines, β – Phenethylamines, Tryptomine, Sterculic acid, Malvalic acid, Linoleic acid.

Medicinal Properties: Root – Aphrodisiac, sweet, tonic.

Leaf – Aphrodisiac, sweet, tonic.

Medicinal Uses: Root – Burning sensation, fever, heart diseases, inflammation, Piles, strangury, urinary complaints.

Leaf – Burning sensation, fever, heart diseases, inflammation, Piles, strangury, urinary complaints.

Ayurvedic preperations: *Baladi kwath, Baladya ghrit, Baladyarista, Chandanbalalakshadi taila, Sudarshan churna, Kukuvadi churna.*

367. *Sida spinosa* L. (Malvaceae)

Distribution: In wastelands & Fallow fields. Hosahangabad, Indore, Mandla, Raisen, Sidhi.

Botany: A woody, much branched herb or undushrub. Stems and branches rough with spiny tubercles at the nodes below the petioles. Leaves ovate to ovate-lanceolate, crenate-serrate glabrous above. Flowers White. Ripe carples 5, toped by 2, erect, rough beaks.

Flowering & Fruiting: October-December.

Medicinal Parts: Whole plant.

Chemical Constituents: Alkaloid as β-Phenethylamine, Ephedrine, Vasicinol, Vasicinone, Vasicine, Choline, Hypaphorine, Methyl ether, Hypaphorine & Betaine.

Medicinal Properties: Root – Antiperiodic, Aphrodisiac, Diaphoretic, Tonic.

Leaves – Emollient, Refrigerant.

Fruits – Astringent, Cooling.

Medicinal Uses: Root – Fevers.

Leaves – Gleet, Gonorrhoea.

368. *Sisymbrium irio* L. (Brassicaceae)

Distribution: In moist places. Bhopal (Oomachan, 1977).

Botany: Annual plant with Yellow flowers in Raceme inflorescence and long straight fruits.

Flowering & Fruiting: December-March.

Medicinal Parts: Seed.

Chemical Constituents: Seed contain Seed yield a semi drying oil.

Medicinal Properties: Seed – Expectorant, Febrifuge, Rubifacient.

Medicinal Uses: Seed – Asthma.

Distribution: Rare. Along the forest margins. Hoshangabad.

369. *Solanum dulcamara* L. (Solanaceae)

Botany: Perennial woody climber, with Purple flowers & Berries fruits.

Flowering & Fruiting: August-January.

Medicinal Parts: Fruits, Stem, Twigs, Root.

Chemical Constituents: The aerial parts contain many alkaloids as Tomatidenol, Solasodine, Solodulicidine, α & β-Solamarine, Solasonine, Soladulcine, Solamargine, Cholesterol, Cycloartanol, Campesterol, Stigmasterol, β-sitosterol.

Fruit contains Yamogenin, β-Solamargine, tigogenin, Diosgenin, Tomatidenol, Cycloartenol, Lanosterol, Cycloartanol, Lophenol, Cycloeucalenol, Obtusifoliol, Gramisterol, Citrostadienol, Cholesterol, Campesterol, Isofucosterol, Sitosterol, Stigmasterol.

Medicinal Properties: Fruits – Alterative, Antirheumatic, Antisyphilitic, Diuretic, Expectorant, Stimulant.

Twigs – Alterative, Antirheumatic, Diuretic, Resolvent.

Medicinal Uses: Fruits – Asthma, Bronchitis, Eczema, Gout, Liver disorders, Psoriasis, Rheumatism, Skin diseases, Syphilis, Warts.

Twigs – Asthma, Psoriasis, Rheumatism, Scrofula, Skin diseases, Ulcers, Urinary troubles, Whooping Cough.

370. *Solanum indicum* L. (Solanaceae)

Syn.: *S. violaceum* Ortega.; *S. sodomenum* L. non ambig.; *S. anguivi* Lam.

Vern.: Bruhati.

Distribution: In Sal and mixed forests. Balaghat, E. Nimar, Hosahangabad, Satna, Seoni.

Botany: Erect, stout, stellately hairy, armed perennial undershrub; leaves ovate-oblong, subacute or obtuse apex, stellately pubescent. Flowers purple in lateral, cymose corymbose; berries globose, orange yellow.

Flowering & Fruiting: July – December.

Medicinal Parts: Fruits.

Chemical constituents: Fruits contain Glycoalkaloids (Solasonine), Solanine, Diosgenin, Carbohydrates, Maltose, saccharase, melibase, solamargine, solasodine, tometidenol, fatty acids.

Leaves contain Solasonine, diosgenin, β – sitosterol, solamargine, solasodine.

Medicinal Properties: Root – Anthelmintic, astringent, bitter, digestive, pungent.

Fruits – Anthelmintic, bitter, pungent.

Medicinal Uses: Root – Leucoderma, fever, asthma, bronchitis, vomiting, pruritus.

Fruits – Pruritus, leucoderma, bronchitis, fever, vomiting, eye diseases.

371. *Solanum nigrum* L. (Solanaceae)

Syn.: *S. subrum* Mill. Ex Wight.; *S. incertum* Dunal ex Grahum.

Vern.: Makoi.

Distribution: In wastelands and moist places throughout the State. Chhatarpur, Damoh, Hosahangabad.

Botany: Unarmed, glabrous or pubescent herbs, with ovate-oblong leaves, white flowers and globose, yellow, red or purple fruits.

Flowering & Fruiting: September– February.

Medicinal Parts: Whole plant, flowers, leaves, fruits.

Chemical constituents: Plant contains Alkaloids, Reducing sugars, glycosides, saponins, steroids, leutein, Lycopene, Vitamin – C, glucose, fructose, caffeic – solasodine, tomatidenel, solamargine, solasomine, trigogenine, potassium, Iron, sulphur, Calcium, Phosphorus.

Medicinal Properties: Fruits – Alterative, aphrodisiac, appetiser, bitter, diuretic, laxative, pungent.

Medicinal Uses: Fruits – Asthma, bronchitis, cough, dysentery, erysipelas, eye diseases, fever, heart diseases, hiccough, inflammation, itching, leucoderma, piles, urinary discharges, vomiting, worms in the ear.

372. *Solanum surattense Burm.* f. (Solanaceae)

Syn.: *S. maccanni* Sant.; *S. xanthocarpum* Schrad. & Wendl.; *S. diffusum* Roxb.; *S. armatum* Br.; *S. virginianum* Jacq.

Vern.: Bhuri rindi.

Distribution: In open wastelands throughout the State. Balaghat, Chhatarpur, Damoh, Dhar, E. Nimar, Hosahangabad, Indore, Mandla, Panna, Rewa, Satna, Seoni, Sidhi, Tikamgarh; W. Nimar.

Botany: Prostrate, prickly herbs, with sinuately lobed to subpinnatifid leaves. Flowers violet-purple or white. Fruits globose, yellow when ripe.

Flowering & Fruiting: December– June.

Medicinal Parts: Whole plant, fruits, seeds.

Chemical constituents: Plant contains Alkaloids (Solasonine), diosgenin. Solacinine.

Seed contains Essential oil, solanocarpine.

Flower contains Flavonol, glycoside, queretin, apigenin, sitosterol.

Leaves contain Coumarine, scopolin, esculin, esculetin.

Medicinal Properties: Root – Anthelmintic, appetiser, bitter, laxative, pungent, stomachic.

Fruits – Anaphrodisiac, anthelmintic, appetiser, bitter, digestive.

Medicinal Uses: Root – Asthma, bronchitis, fever, heart diseases, lumbago, pains, piles, strangury.

Fruits – Asthma, fever, heart diseases, pruritus.

Ayurvedic preperations: *Dashmula ghrit, Nidigdhadi kwath, Vyaghri taila, Vyaghriharitaki, Dashmularishta, Dashmula taila, Dashmula kwath, Kantkari ghrit, Kantkari avleha, Sudarshan churna, Sarvahara louh, Kanakasava, Devadarvadi kwath, Sringayadi churna.*

373. *Soymida febrifuga A. Juss.* (Meliaceae)

Syn.: Swietenia febrifuga Roxb.

Vern.: Rohan.

Distribution: Jabalpur.

Botany: A tall glabrous deciduous trees; flowers Greenish-White, in terminal or axillary panicles.

Flowering & Fruiting: March-May.

Medicinal Parts: Bark.

Chemical constituents: Bark contains Bitter resinous substance, tannic acid, galic acid.

Medicinal Properties: Aphrodisiac, astringent, bitter, cooling, febrifuge, nutritive, wound healing.

Medicinal Uses: Chronic fever, diarrhoea, dysentery, malaria fever, skin diseases, throat diseases, wounds.

374. *Spermacoce hispida* L. (Rubiaceae)

Syn.: *Borreria hispida* (L.) K. Schum.

Distribution: Balaghat, Bhopal, Damoh, Hosahangabad, Indore, Jabalpur, Mandla, Narsimhpur, Panna, Satna, Seoni, Shahdol, Sidhi.

Botany: Erect or diffuse hispid herbs; flowers Pink; fruits ellipsoid capsules.

Flowering & Fruiting: August-April.

Medicinal parts: Whole plant

Chemical constituents: Alkaloid as Borreline; β-Sitosterol, Urosolic acid, Iso-rhamnetin.

Medicinal properties: Acrid, Astringent, Emollient, Stomachic, Sweet.

Medicinal uses: Colic, Dyspepsia, Flatulence, General debility.

375. Sphaeranthus indicus L. (Asteraceae)

Syn.: *Sphaeranthus mollis* Roxb. ex DC.; *S. hirtus* Willd.

Distribution: Common in dry open forests. Balaghat, Damoh, Hosahangabad.

Botany – Prostrate or decumbent-ascending, aromatic annual herb with Purple flowers & Achenes fruits.

Flowering & Fruiting: November-March.

Medicinal parts: Whole plant.

Chemical constituents: Essential oil contains – Methyl Chavicol, α-Ionone, d-Cadinene, p-Methoxycinnamaldehyde, α-Terpinene, Citral, Geraniol, Geranyl acetate, β-Ionane, Sphaerene, Indicusene, Sphaeranthol.

Medicinal properties: *Root* & *Seed* – Anthelmintic, Stomachic.

Flowers – Alternative, Cooling, Tonic, Depurative, Refrigerants.

Medicinal uses: *Flowers* – Conjunctivitis, Skin diseases.

Root & *Seed* – Intestinal worms, Indigestion, Cough.

376. *Sterospermum chelonoides* DC. (Bignoniaceae)

Syn.: S. personatum (Hassk.) D. Chatterjee.

Vern.: *Padar.*

Distribution: Common along forest roads. Balaghat, Jabalpur, Mandla, Seoni.

Botany: Large deciduous trees, with Yellow or crimson flowers tinged into Red colour and cylindric capsules.

Flowering & Fruiting: April-January.

Medicinal Parts: Root, Leaves, Flowers, Fruits, Seeds.

Chemical Constituents: Leaves – Flavone as Stereolensin; Lapachol, Dinatin-7-Glucuronide, 6-O-Glucosylcutellarein, 6-O-Glucosylcutellarein.

Root bark – β-Sitosterol, n-Triacontanol, Specioside.

Heartwood – Lapachol, Dehydro-a-Lapachone, Dehydrotectol.

Medicinal Properties: Root – Antiasthamatic, Anticancer, Anti-Inflammatory, Antitumour, Antiviral, Astringent, Cardiotonic, Cooling, Diuretic, Expectorant, Febrifuge, Tonic.

Leaves – Antimalarial, Antirheumatic, Antitumour.

Flowers – Aphrodisiac, Astringent, Cardiotonic, Sweet, Tonic.

Fruits – Antileprotic, Astringent, Cooling, Sweet.

Medicinal Uses: Roots – Anorexia, Asthma, Cough, Diarrhoea, Dyspepsia, Fever, General debility, Haemorrhoids, Hiccough, Piles.

Flowers – Bleeding, Cardiopathy, Burning sensation, General debility, Hiccough.

Fruits – Blood diseases, Hiccough, Leprosy, Strangury.

377. *Sterospermum colais* (Buch. – Ham. ex Dillw.) Mabberley (Bignoniaceae)

Syn.: *S. personatum* (Hassk.) Chatterjee

Vern.: *Chhota padar.*

Distribution: In mixed forests or hill slopes. Seoni.

Botany: Large deciduous trees, with Yellow flower & Capsules fruits.

Flowering & Fruiting: May-August.

Medicinal Parts: Roots, Leaves, Fruits, Seeds.

Chemical Constituents: Leaves – A flavone as Stereolensin & Lapachol.

Root – Lapachol, β-Sitosterol.

Medicinal Properties: Root – Acrid, Anodyne, Anti-Inflammatory, Aphrodisiac, Appetiser, Astringent, Bitter, Constipating, Diuretic, Expectorant, Febrifuge, Toinc.

Flowers – Aphrodisiac, Astringent, Cooling, Sweet, Tonic.

Fruits – Astringent, Cooling, Sweet.

Medicinal Uses: Roots – Asthma, Cough, Diarrhoea, Dyspepsia, Hiccough, Hyperacidity, Hyperdipsia, Renal and Vesical calculi, Strangury.

Leaves – Cooling, Malaria, Otalgia, Wounds.

Flower – Burning sensation, Hiccough, General debility.

378. *Strychnos potatorum* L. f. (Loganiaceae)

Vern.: *Kaya.*

Distribution: In mixed forests. Hosahangabad, Seoni.

Distribution: West Bengal, Central & South India upto 1,200 m.

Botany: Medium sized deciduous glabrous trees, with fragrant White flowers & berries fruits.

Flowering & Fruiting: April-July.

Medicinal Parts: Seeds.

Chemical Constituents: Seeds – Brucine, Diaboline, β-Sitosterol, Stigmasterol, Oleanolic acid and its 3B-acetate and a Saponin, Galactose and Mannose from Seeds.

Leaf & Bark – Isomotiol, Mixture of Sitosterol, Stigmasterol, Campsterol.

Medicinal Properties: Seed – Alexeteric, Anthelmintic, Aphrodisiac, Appetiser, Astrignent, Bitter, Digestive, Dimulcent, Diuretic, Emetic, Refrigerant, Sweet, Tonic.

Medicinal Uses: Seed – Bronchitis, Burning sensation, Chronic diarrhoea, Conjunctivitis, Diabetes, Dipsia, Dysentery, Gastropathy, Gonorrhoea, Leucorrhoea, Strangury, Ulcers.

Root – Leprosy.

379. *Symplocos racemosa Roxb.* (Symplocaceae)

Vern.: Lodhra.

Distribution: Common in deciduous forests. Satna, Sidhi.

Botany: A small tree about 20 ft. high with greyish bark; Leaves elliptic-ovate, crenate, and glabrous; Flowers aromatic, yellow; Drupe ellipsoid.

Flowering & Fruiting: September-May.

Medicinal Parts: Stem bark.

Chemical constituents: Bark contains Alkaloids: Leupeorin, Leupeoridin, Glycosides, Loturin, Loturidin, Colloturine.

Medicinal Properties: Anti-inflammatory, antipyretic, astringent, blood purifier, cooling, styptic, tonic.

Medicinal Uses: Diarrhoea, dysentery, elepantiasis, eye troubles, fever, leprosy, leucorrhoea, liver troubles, piles, ulcer.

Ayurvedic preperations: *Lodhrasav, Lodhradi kwath, Pippalayasav.*

380. *Syzygium* cumini (L.) Skeels (Myrtaceae)

Syn.: *Eugenia jambolana* Lam.

Vern.: Jamun.

Distribution: Growing in sal forest margins, along road sides, streams and rivers. Balaghat, Damoh, Hosahangabad, Mandla, Shahdol, Shivpuri, Sidhi.

Botany: Trees, with elliptic-lanceolate leaves and pale-green or white flowers in axillary paniculate cymes. Fruits globose or oblong, purple when ripe.

Flowering & Fruiting: February-August.

Medicinal Parts: Leaves, bark, fruits, seed.

Chemical constituents: Bark contains Friedelin, kaempferol, tannins, quercetin, β – sitosterol, betullinic acid.

Fruits contains Anthocyanin, eugin.

Seed contains Ellagic acid, IAA, Oxalic acid, citric acid, glycollic acid, glucose, fructose, gallic acid.

Leaves contain Glycine, alanin, leucin, tyrosin.

Medicinal Properties: Bark – Acrid, anthelmintic, digestive, sweet.

Fruits – Acrid, astringent, cooling, sweet.

Seed – Astringent, sweet.

Medicinal Uses: Bark – Asthma, biliousness, blood diseases, bronchitis, dysentery, Sore throat, ulcer.

Fruits – Biliousness.

Seed – Diabetes.

Ayurvedic preperations: *Panchpallav yog, Jambuphalasav, Nyagrodhadi churna, Jambadhya taila, Jambvarista.*

381. *Tabernaemontana divaricata* (L.) R. Br. ex Roem. & Schult. (Apocynaceae)

Syn.: *E. coronaria* (Jacq.) Stapf.; *Ervatamia divaricata* (L.) Burkill.; *T. coronaria* (Jacq.) Willd.

Vern.: *Chandni.*

Distribution: Ornamental. Balaghat, Chhatarpur, Damoh, Dhar, Indore, Shahdol.

Botany: Glabrous shrubs or small trees, with oblong or obovate-lanceolate, acuminate leaves; flowers white, with about 5 cm. long follicles.

Flowering & Fruiting: May-August.

Medicinal Parts: Root, Root-bark, Leaves, Stem-bark, Flowers, Wood, Milky juice.

Chemical constituents: Seed contains Citric, oleic, palmitic acid.

Leaves, Stem/Roots contain Dregamine, epiervatamine, tabernaemontaine, vobasine, voaconjine, voacamine, flavonoid, flavonol.

Flowers contain Janetine, hecubine, kampferol.

Stem bark contains α-Amyrin, lupeol, β-sitosterol, coronaridine, apparicine, ervaticine, ervatinine.

Latex contains Enzymes, amino acids, proteins, galactose.

Medicinal Properties: Root – Acrid, bitter, thermogenic, anodyne, astringent, vermifuge.

Flowers – Cooling, fragrant.

Latex – Cooling, fragrant.

Medicinal Uses: Root – Paralysis, strangury.

Flowers – Burning sensation.

Latex – Wounds.

382. *Tabernaemontana heyneana Wall.* (Apocynaceae)

Syn.: *T. alternifolia* L.; *Ervatamia heyneana* (Wall.) Cooke.; *Ervatamia alternifolia* (L.) Almedia.

Distribution: Rare. Mandala.

Botany: A shrub or small deciduous tree.

Flowering & Fruiting: May-August.

Medicinal Parts: Wood, Milky juice, Root & Stem bark.

Chemical Constituents: Root – Coronaridine, Voacangine, Ibogamine.

Fruits – Lupeol-O Ac and β-Amyrin-OAC.

Leaves – Alkaloid-Isocoacristine.

Medicinal Properties: Root – Acrid, Alexipharmic, Astringent, Bitter, Digestive.

Medicinal Uses: Root – Biliousness, Diseases of blood.

383. Tamarindus indica L. (Caesalpiniaceae)

Syn.: *T. officinalis* Hook.; *T. occidentalis* Gaertn.

Vern.: Imli.

Distribution: Balaghat, Bhopal, Damoh, Indore, Jabalpur, Mandla, Raisen, Shahdol, Shivpuri, Sidhi.

Botany: Evergreen trees, with 6-12 pairs of leaflets and yellow with purple streaks. Pods subtorulose, falcate.

Flowering & Fruiting: May-August.

Medicinal Parts: Leaves, stem bark, fruits, seeds, flowers.

Chemical constituents: Polysaccharides, balsamine, catechin, nasturtium, tamarin, phosphatidic acid and its choline, ethanollamine, serine, inositol, alkaloid, citric acid, tartaric acid, potassium bitartrate.

Medicinal Properties: Bark – Astringent.

Leaf – Bitter, laxative, sweet.

Flower – Acrid, appetiser, sour, sweet.

Fruits – Anthelmintic, appetiser, sour, sweet.

Medicinal Uses: Bark – Gonorrhoea, paralysis, urinary discharges.

Leaf – Blood diseases, earaehe, eye diseases, inflammation, ophthamia, ringworm, small pox, snake bite, swelling, tumor.

Flower – Urinary discharges.

Fruits – Wound fractures.

Seed – Ulcers, vaginal discharges.

Ayurvedic preperations: *Cinca ksara, Cinca panaka, Sankha vati.*

384. *Tectona grandis* L. f. (Verbenaceae)

Vern.: Sagwan.

Distribution: Balaghat, Chhatarpur, Dhar, Hosahangabad, Panna, Rewa, Seoni, Tikamgarh.

Botany: Deciduous trees, with broadly elliptic large leaves. Flowers white or pinkish, stamens all equal. Fruits enclosed in bladder-like inflated calyx.

Flowering & Fruiting: September–December.

Medicinal Parts: Whole plant, bark, wood, leaves, flowers, seeds.

Chemical constituents: Tactograndon, Betulin, Lupeol, Betulinic acid, Betulinaldehyde, Carbohydrates, fatty acids, lapachanone, deoxylapachol, tectoquinine.

Medicinal Properties: Flower – Acrid, bitter, dry.

Bark – Acrid, sweet.

Wood – Acrid, cooling, laxative.

Medicinal Uses: Flower – Biliousness, bronchitis, urinary discharges.

Bark – Bronchitis.

Wood – Biliousness, dysentery, leucoderma, piles.

385. *Tephrosia purpurea Pers.* (Fabaceae)

Syn.: *T. Villosa* Pers.; *T. maxima* Pers.; *T. lanceolata* R. Grah., *T. hamiltonaii* Drumm ex Gamble.

Vern.: Sarphonka.

Distribution: Balaghat, Bhopal, Bhind, Dhar, Hosahangabad, Indore, Jabalpur, Mandla, Morena, Narsimhapur, Panna, Raisen, Rewa, Sagar, Satna, Seoni, Shahdol, Shivpuri, Sidhi, Tikamgarh.

Botany: Perennial, woody herbs, with 11-21 leaflets and reddish-purple or bright pink to white flowers in leaf-opposed racemes. Style glabrous.Pods linear.

Flowering & Fruiting: Throughout the year.

Medicinal Parts: Whole plant, root, Root bark, seeds.

Chemical constituents: Beta – Sitosterol, Cyanidin Chloride, Pongomol, Lupeol, Rutin, Dephinindinchloride, Karangin, Kangone, flevichapparins – B and C, Methylkaranjic acid, Purgurin.

Root contains Tephrosin, Degulin, Rotenone, Isotephrosin, alterative.

Medicinal Properties: Acrid, alexeteric, alterative anthelmintic, antipyretic, bitter, digestive.

Medicinal Uses: Diseases of blood, heart, liver, spleen. asthma, bronchitis, leprosy, snake bite, tumour, ulcer, wound.

Ayurvedic preperations: *Sarpunkha svarasa, Sarpunkha lepa, Sarpunkha Ghana vati.*

386. Teramnus labialis (L. f.) Spr. (Fabaceae)

Vern.: Mashoni.

Distribution: Climbing on roadsides bushes. Balaghat, Damoh, Gwalior, Hosahangabad, Indore, Jabalpur, Mandla, Narsimhapur, Panna, Raisen, Rewa, Sagar, Seoni, Shahdol, Shivpuri, Sidhi.

Botany: A climbing wide spreading hairy herb; leaves trifoliate, leaflets are of unequal size; flowers pink or bluish white.

Flowering & Fruiting: August-April.

Medicinal Parts: Leaves, Fruits, Seeds.

Chemical constituents: Galactomannan, carbohydrates, sugars, amino acids, fatty acids, carotene, saponins, sapogenin, diosgenin, gitogenin, neogitogenin; Alkaloids, Saponeritin, neogigenin, tigogecnin, vicenin, vitexin, luteolin, yamogenin, furastanolglycosides, kaempfrol, quercetin.

Medicinal Properties: Seeds – Aphrodisiac, astringent, carminative, demulcent, diuretic, tonic.

Medicinal Uses: Seeds – Anaemia, colic, cough, diabetes, diarrhoea, dropsy, dysentery, rheumatism, rickets.

387. *Terminalia arjuna* (Roxb.) Wt. & Arn. (Combretaceae)

Vern.: Arjun,

Distribution: Along banks of streams and rivers, and in mixed forests. Balaghat, Bhopal, Chhatarpur, Chhindwara, Damoh, Hosahangabad, Indore, Jabalpur, Mandla, Raisen, Rewa, Sagar, Satna, Seoni, Shahdol, Shivpuri, Sidhi, Tikamgarh.

Botany: Trees, with smooth bark and leaves cordate at base. Flowers white, in axillary spikes or terminal panicles. Fruits 5-winged, wings curved upwards.

Flowering: April-May.

Fruiting: Almost throughout the year.

Medicinal Parts: Bark, leaves, fruits.

Chemical constituents: Bark contains Bita – sitosterol, Arjunic acid, Friedlene, Glucoside, (Arjunetic), Tannins, Sugars, Sodium, Magnessium, Aluminium, Calcium Carbonate.

Medicinal Properties: Bark – Acrid, alexiteric, antidysenteric, cooling, heating, styptic, sweet, tonic.

Medicinal Uses: Bark – Asthma, biliousness, blood diseases, fractures, leucoderma, tumour, ulcer, urinary discharges.

Ayurvedic preperations: *Arjunarista, Arjuna ghrit, Parthadhyarista, Arjunatwak churna, Kukubhadi churna, Hridayarnava ras and Arjun Ksheerpak, Aravindasava, Devadarvyarista.*

388. *Terminalia bellerica* (Gaentn.) Roxb. (Combretaceae)

Vern.: Bahera.

Distribution: In mixed and Sal forests. Balaghat, Bhopal, Damoh Hosahangabad, Indore, Jabalpur, Mandla, Raisen, Rewa, Sagar, Satna, Seoni, Shahdol, Shivpuri, Sidhi.

Botany: Deciduous trees, with broadly obovate leaves clustered at the end of branches. Flowers yellow or creamy white, in spikes. Fruits ellipsoid, 5-ridged.

Flowering: April-May.

Fruiting: Most part of the year.

Medicinal Parts: Fruits,

Chemical constituents: Fruits contain β – Sitosterol, gallic acid, ellagic acid, Chebulagic acid, Galloyl, Glucose, Many free sugars, Manitol, galactose, Rhamnose, glucoside.

Medicinal Properties: Fruits – Acrid, anthelmintic, bitter, digestive, laxative, Pungent.

Medicinal Uses: Bark – Anaemia, leucoderma

Fruits – Asthma, biliousness, bronchitis, cardiac diseases, inflammation, sore throat.

Ayurvedic preperations: *Bibhitaki sura, Bibhitaka ghrit/Avleha/Churna, Bibhitakadi kwath, Bibhitaka taila, Triphala churna, Phalatrikadi kwath, Talisadi churna, Lavangadi vati, Sudarsana churna, Sarvajvar louha, Agnitundi vati, Sanjivani vati, Punarnava mandur, Navayas louha, Sringadi churna, Karanjadi yog, Laghuvis-garva taila, Arogyavardhini vati, Khadirarista, Gokshuradi guggul, Chandraprabha vati, Vidangadi louha, Triphala ghrit/kwath, Balaguti.*

389. *Terminellia chebula* (Gaentn.) Retz. (Combretaceae)

Vern.: Harra

Distribution: In dry deciduous and sal forests. Balaghat, Betul, Chhindwada, Damoh, Hosahangabad, Indore, Jabalpur, Mandla, Raisen, Seoni, Shahdol, Sidhi.

Botany: Deciduous trees, Leaves ovate or elliptic. Flowers Yellow or creamy white, in spikes. Fruits 5-ridged.

Flowering: April-June.

Fruiting: Winter Season.

Medicinal Parts: Fruits.

Chemical constituents: Fruits contain – Tannic acid, gallic acid, chebulinic acid, mucilage, glycoside (Anthroquinine), Carbohydrates, amino acids, phosphoric acid, succinic acid.

Medicinal Properties: Fruits – Alterative anthelmintic, antidysenteric, carminative, expectorant, heating, stomachic, tonic.

Medicinal Uses: Fruits – Anaemia, ascities, asthma, biliousness, constipation, elephantiasis, gout, hiccough, inflammation, leucoderma, piles, sore throat, strangury, tumour, typhoid, urinary discharges, vesicular calculi, vomiting.

Ayurvedic preperations: *Abhaya modaka, Abhyarista, Agastiharitaki, Agnitundi vati, Amrita haritaki, Amlakyadi churna, Arogyavardhini vati, Aragbadhyadi kwath, Byaghri haritaki, Chittrak haritaki, Chandraprabha vati, Gokshuradi guggul, Danti haritaki, Haritaki khand, Haritakadi kwath,*

Khadirarista, Kukuvadi churna, Karanjadi yog, Loghu-vis-garva-taila, Navayas louha, Parthyadi kwath, Parthayadi churna, Panchaskar churna, Phaltrikadi kwath, Punarnava mandur, Sanjivani vati, Sarvajvarhar louha, Sringayadi churna, Sudarsana churna, Dashmula haritaki, Triphala, Vasadi kwath, Vidangali lep.

390. *Terminalia crenulata Roth. (*Combretaceae)

Vern.: Saj.

Distribution: In deciduous forests, also planted along the roadsides. Bhopal, Raisen.

Botany: A medium sized to large tree, with simple leathery leaves and winged fruits.

Flowering & Fruiting: Most part of the year.

Medicinal Parts: Bark.

Chemical constituents: Bark – B-sitosterol, arjunic acid, betulinic acid, ellagic acid, tannin.

Medicinal Properties: Bark – Anthelmintic, astringent, bitter, cardiotonic, constipating, cooling, diuretic, expectorant, styptic, vulnerary.

Medicinal Uses: Bark – Bronchitis, burning sensation, cough, diarrhoea, dysentery, gonorrhoea, haemorrhages, leucorrhoea, ulcers.

391. *Thalictrum foliolosum* DC. (Ranunculaceae)

Vern.: Pilzari.

Distribution: In moist shades places on higher hill slopes. Hosahangabad, Shahdol, Sidhi.

Botany: Perennial herbs, with Green or Violet flowers & Achene fruits.

Flowering & Fruiting: July-September.

Medicinal Parts: Roots.

Medicinal Properties: Root – Diuretic, Purgative, Bitter.

Medicinal Uses: Conjunctivitis, Leucoderma.

Root – Ophthalmia, Dyspepsia.

392. *Thespesia populnea* (L.) Soland. ex Correa. (Malvaceae)

Syn.: *Hibiscus populnea* **L.**

Distribution: Commonly planted as an avenue tree. Bhopal, Dhar.

Botany: Small evergreen trees, with yellow flowers and brown capsule fruits.

Flowering & Fruiting: September-August.

Medicinal Parts: Bark, stem, leaves, root, fruit.

Chemical constituents: Leaves contain Lupeol, lupenone, B-sitosterol.

Stem contains Calycopterin, sequiterpenoid quinones.

Fruits contains Thespesin, gossypol, populinin, populneol, herbacetin, populnetin, gossypectin, epoxyleic acid, isoquercitrin, rutin, B-sitosterol, thespesone.

Medicinal Properties: Bark – Alterative, antidiabetic, antidysenteric, astringent, cooling, diuretic.

Leaves – Anti-inflammatory, antipyretic, astringent, cephalalgic.

Root – Anti-inflammatory, antiscorbutic, antiseptic, sweet.

Fruit – anticancer, antifertility, antimalaria, antitumour, antiviral.

Medicinal Uses: Bark – Cough, diabetes, dysentery, dyspepsia, gonorrhoea, headache, itching, influenza, indigestion, psoriasis, typhoid, ulcers.

Leaves – Cough, eczema, fever, headache, inflammations, influenza, psoriasis, scabies.

Root – Inflammations, psoriasis, scabies.

Fruit – Cancer, eczema, malaria, migrain, scabies, psoriasis, ringworm, tumours.

393. *Thevetia peruviana* (Pers.) Schum. (Apocynaceae)

Syn.: *T. nerifolia* Juss. ex Steud.; *Cascabela thevetia* (L.) Lipp.

Vern.: Pili kaner.

Distribution: Cultivated in gardens and parks. Balaghat, Chhatarpur, Damoh, Indore, Panna, Sidhi.

Botany: A large evergreen shrub or small trees, with linear leaves, yellow flowers and subglobose fruits.

Flowering & Fruiting: Throughout the year.

Medicinal Parts: Root, Leaves, Bark, Kernel, Seeds, Latex.

Chemical constituents: Seeds contain – Theveside, viridoside, thevefolin, thevetin, thevetoxin, peruvoside, ruvoside.

Flowers & Fruits contain – Epiperuviol acetate, α – & β – amyrin, kaempferol, quercetin, glycoside – thevetin & other thevetoxin.

Medicinal Properties: Antipyretic, cardiotonic, diuretic, febrifuge.

Medicinal Uses: Bark – Dropsy, oedema, rheumatism.

394. *Tinospora cordifolia* (Willd.) Miers. ex Hk. f. & Th. (Menispermaceae)

Syn.: *T. glabra* (N. Burm.) Merr.; *Cocculus cordifolius* DC.; *Menispermum cordifolium* Willd.

Vern.: Gurcha.

Distribution: On tree in the forests. Bhopal, Indore, Rewa, Sagar, Satna.

Botany: Succulent, glabrous, deciduous climbers, with corky and lenticelled bark. Leaves ovate-cordate. Flowers yellow, in axillary racemes. Fruits globose, red when ripe.

Flowering & Fruiting: August-February.

Medicinal Parts: Whole plant, Juice of fresh plant.

Chemical constituents: Starch, Glycoside (Giloin), Tinosporic acid, Vasa alcohol, essential oil, Vasa acid, Proteins, Calcium, Phosphorus, Berberine, gilosterol.

Medicinal Properties: Stem – Bitter, diuretic, stimulant, stomachic.

Medicinal Uses: Stem – Burning sensation, constipation, diabetes, enriches the blood, fevers, Jaundice, skin diseases, spleen diseases tonic, vaginal discharges, vomiting.

Ayurvedic preperations: *Guduchi sattva, Amrita guggula, Guducyadi kasaya-taila and churna, Guduchayadi kwath, Guduchayadi churna, Suduchi louha, Amritarista, Kirattikadi kwath, Sudarshan churna, Sanjivani vati, Amritastak kwath, Phalatrikadi kwath, Guduchi taila, Pippalimuladi kwath, Panchbhadra kwath, Rasnasaptak kwath, Yograj guggul, Kaisur guggul, Rasnapanchak kwath, Sansmani vati, Arkharabhara ras, Sat giloe.*

395. *Toona ciliata Roem.* (Meliaceae)

Syn.: *T. hexandra* (Wall. ex Roxb.) *Roemer.; Cedrala toona* Roxb.

Vern.: Tuna.

Distribution: Hosahangabad, Mandla.

Botany: A large deciduous tree; Leaves paripinnate; leaflets 8-30, lanceolate or ovate-lanceolate, entire or slightly undulate, often oblique at the base; flowers white, fragrant, in drooping, terminal panicles; capsule oblong, smooth, dark brown, septifragally 5-valved; seeds reddish-brown, winged at both ends.

Flowering & Fruiting: March – April.

Medicinal Parts: Stem bark, flowers.

Chemical constituents: Wood & seeds contain A bitter substance – Nyctanthin, cedralone, β-sitosterol.

Flowers contain Baragapten flavone, glucoside, nyctanthin.

Bark contain Toonacilin, toonacilids, bitter, resin.

Wood contain Ferneslene, geronyl, geraniol, siderin, volatile oil.

Medicinal Properties: Bark – Acrid, anthelmintic, aphrodisiac, astringent, bitter, digestive, expectorant, fattening, sweet, tonic.

Medicinal Uses: Bark – Biliousness, blood diseases, burning sensation, fever, headache, itching, leprosy, scabies, ulcer.

396. *Trachyspermum ammi* (L.) Spragae. (Apiaceae)

Syn.: *Carum copticum* Benth & Hook.

Vern.: Ajwain.

Distribution: Cultivated. Damoh, Hosahangabad.

Botany: Annual herbs, with whitish or pinkish flowers and muricated fruits.

Flowering & Fruiting: November-March.

Medicinal Parts: Fruits, Leaves, Oil.

Chemical Constituents: Fruits contain – An aromatic Volatile oil, Stearoptene, Cumene, Terpene as Thymine; Phenolic Glucosides; Thymol, p-cymene, Stearoptin, Cymon, Terpene.

Medicinal Properties: Fruits – Anthelmintic, Antidiarrhoeal, Appetiser, Digestive, Stomachic, Tonic.

Leaves – Anthelmintic, Appetiser, Digestive, Heart tonic, Stimulant, Stomachic, Tonic.

Medicinal Uses: Fruits – Abdominal Colic & Pain, Cholera, Diarrhoea, Dyspepsia, Flatulence, Heart tonic, Helminthiasis, Indigestion, Rheumatism, Uritcaria.

Leaves – Abdominal pain, Cholera, Colic, Diarrhoea, Flatulence, Heart diseases, Helminthiasis, Indigestion, Rheumatism, Urticaria.

397. *Trapa bispinosa* (Roxb.) Makino. (Trapaceae)

Syn.: *T. quadrispinosa* Wall.; *T. natans* L. var. bispinosa (Roxb.) Makino.

Vern.: Singhara.

Distribution: Common in lakes, ponds and tanks, sometimes cultivated. Almost throughout the State.

Botany: An aquatic floating herbs; leaves of two kinds; submerged opposite pectinate; floating clustered, rhomboid, red and villous beneath; petioles inflated purple, solitary; fruits 2-horned.

Flowering & Fruiting: August-December.

Medicinal Parts: Seeds.

Chemical constituents: Fruit contains – Citric acid, tannin, amylose, amylopectin, carbohydrates, β-amylase, phosphorylase enzymes, protein, fat, nicotinic acid, riboflavin, thiamine, vitamine A & C, mangenese.

Medicinal Properties: Anti-inflammatory, astringent, refrigerant, stomachic, tonic.

Medicinal Uses: Fruits – Diarrhoea, leucorrhoea, seminal weakness.

398. *Trianthema portulacastrum* L. (Azoiaceae)

Distribution: Common on railway tracts and roadsides.

Botany: Succulent, prostrate herbs, with spathulate leaves. Flowers white or pink, solitary; stamens 10. capsules mitrifrom.

Flowering & Fruiting: **August.**

Medicinal Parts: Roots, Leaves, Fruits, Seeds.

Chemical Constituents: Protein, fat, carbohydrate, crude fibre, calcium, phosphorus, Iron, ascorbic acid, carotene, oxalates.

Medicinal Properties: Plant – The plant is Alexiteric, Analgesic, Alterative, Bitter, Hot, Laxative, Stomachic.

Medicinal Uses: Plant – Anaemia, Ascites, Bronchitis, Heart diseases of the blood, Inflammation, Piles,

Root – Night blindness.

399. *Tribulus terrestris* L. (Zygophyllaceae)

Vern.: Gokhru.

Distribution: Hosahangabad, Indore, Ratlam, Satna, Sidhi, Tikamgarh.

Botany: Procumbent annual herbs, with pinnate leaves and yellow flowers. Mericarps dorsally tuberculate, spines 4.

Flowering & Fruiting: May-June.

Medicinal Parts: Fruits, roots.

Chemical constituents: Fruits contains Alkaloid, fixed oil, essential oil, Resin, Steroids, Saponins, Vitamin – C, Protein, Nitrates.

Medicinal Properties: Root – Alterative, aphrodisiac, appetiser, cooling, fattening, Sweet, tonic.

Fruits – Alterative, aphrodisiac, appetiser, cooling, fattening, Sweet, tonic.

Seed – Aphrodisiac, cooling, diuretic, fattening.

Medicinal Uses: Root – Asthma, blood diseases, burning sensation, cough, inflammation, pruritus, strangury urinary discharges, vesicular calculi.

Fruits – Asthma, blood diseases, burning sensation, cough, inflammation, pruritus, strangury urinary discharges, vesicular calculi.

Seed – Inflammation, stones in bladder, urinary trouble.

Ayurvedic preperations: *Dasamularista, Goksuradi ghrit, Goksura kwath, Dashmula kwath, Gokshuradi churna, mGokshuradi kwath, Gokshuradi guggulu, Gokshuradi avalcha, Abhyarista, Chavyanprash, Rasna-saptak kwath, Haritakayadi kwath, Brihat varunadi kwath.*

400. *Trichodesma indicum* (L.) R. Br. (Boraginaceae)

Vern.: Chhota-Kulpha.

Distribution: Common in wastelands, grassy fields & along roadsides. Dhar, Indore, Seoni.

Botany: An erect or diffuse, much branched, Grey-Villous hispid herb. Leaves mostly sessile lanceolate or cordate-lanceolate. Flowers pale-Blue or Violet or White with Brown throat. Nutlets rough on the inner face, obscurely margined.

Flowering & Fruiting: July-November.

Medicinal Parts: Whole plant.

Medicinal Properties – Plant – Acrid, Alexeteric, Anodyne, Anti-Inflammatory, Bitter, Carminative, Constipating, Depurative, Diretic, Emollient, Febirufuge, Ophthalmic, Thermogenic,

Medicinal Uses: Plant – Diarrhoea, Dysentery, Dysmenorrhoea, Dyspepsia, Fever, Inflammation, Leprosy, Skin diseases, Strangury.

401. *Trichosanthes cucumerina* L. (Cucurbitaceae)

Syn.: *T. anguina* L.

Vern.: Jangli-chichinda.

Distribution: Balaghat, Bhopal, Hosahangabad, Indore, Panna, Sidhi, Tikamgarh

Botany: A climbing or spreading, scabrous, stout herb. Tendrils divided. Leaves usually 5-lobed or angled; Margins denticulate. Flowers White, with fimbriase petals. Male peduncles paired. Female flowers solitary. Fruits Yellow, pointed.

Flowering & Fruiting: June-October.

Medicinal Parts: Whole plant.

Chemical Constituents: Leaves – β-Sitosterol Rits gloucoside and luteolin-7-glucoside.

Seed oil – Punicic acid, Oleic acid, Linoleic acid, Palmitic acid, stearic acid, arachidic acid, Cucurbitacin-B.

Medicinal Properties: Roots / Seeds – Anthelmintic, Cooling, Febrifuge, Purgative, Vermifuge, Tonic.

Leaves – Cooling.

Fruits – Anthelmintic, Carminative, Cooling, Digestive, Febrifuge, Purgative, Sweet, Tonic.

Medicinal Uses: Roots / Seeds – Syphilis, Verminosis.

Fruits – Anorexia, Burning sensation, Constipation, Dipsia, Dyspepsia, Fever, Flatulence, General weakness.Helminthiasis.

402. *Tridax procumbens* L. (Asteraceae)

Distribution: Common in waste moist places. Balaghat, Hosahangabad, Satna.

Botany: Straggling herbs, with ovate-lanceolate leaves & yellow heads terminating long peduncles. Achenes silky.

Flowering & Fruiting: Throughout the year.

Medicinal Parts: Whole plant.

Chemical Constituents: Leaf – Fumaric acid.

Flowers – Glucoluteolin, Luteolin, Isoquercetin, Quercetin.

***Medicinal Properties*:** Plant – Hypotensive.

Leaf juice – Insecticidal.

Medicinal Uses: Wounds, Boils.

403. *Trigonella foenum-graceum* L. (Fabaceae)

Syn.: *Trifolium foenum-graceum*; *Foenum-graceum officinale.*

Vern.: Methi.

Distribution: Cultivated for seeds, used as a spice. Bhopal, Shivpuri.

Botany: An aromatic, erect, annual herbs with 3-foliate leaves and pale yellow flowers in axils of young leaves; pods 8-10 cm. long, with long beak.

Flowering & Fruiting: February.

Medicinal Parts: Seeds.

Chemical constituents: *Seed contains* – Quercetin, luteolin, trigonelline, choline, nicotinic acid, fixed oil, essential oil, mucilage, bitter extractive, colouring matter, 2 steroidal saponins.

Medicinal Properties: Seeds – Anti-inflammatory, anaphrodisiac, aromatic, astringent, bitter, carminative, diuretic, emollient, stomachic, thermogenic, tonic.

Medicinal Uses: Seeds – Bronchitis, cough, dropsy, epilepsy, fever, gout, paralysis, vomiting.

404. *Triticum aestivum* L. (Poaceae)

Vern.: **Genhu.**

Distribution: Cultivated throughout the state.

Botany: Medium-sized annuals. Inflorescence a terminal distichous spike; Spikelets solitary, awned. Grains elliptic-oblong.

Flowering & Fruiting: January-April.

Medicinal Parts: Grains.

Chemical Constituents: Grians – Protein, Crude fibres, Sugars, Fat, Minerals, Vitamins, Enzymes.

Medicinal Properties: Grains – Aphrodisiac, Appetiser, Demulcent, Emollient, Laxative,

Refrigerant, Rejuvenating, Sweet, Vulnerary.

Medicinal Uses: Grains – Anorexia, Boils, Constipation, Flatulence, General debility, Mernorrhagia, Pruritus, Skin diseases, Ulcers.

405. *Triumfetta rhomboidea Jacq.* (Tiliaceae)

Syn.: *T. bartramia* L.; *T. angulata* Lam.

Distribution: Along the forest margins & roadsides. Hosahangabad, Jabalpur, Mandla, Sagar, Seoni, Shahdol, Sidhi.

Botany: Stout herbs, with ovate-rounded, 3-lobed & entire Leaves. Flowers yellow, with 5-10 stamens. Capsules pubescent, spines glabrous.

Flowering & Fruiting: August-November.

Medicinal Parts: Seeds, Bark.

Chemical Constituents: Plant contains – 4-Hydroxyisoxazole and Triumboidin.

Medicinal Properties: Root – Acrid, Aphrodisiac, Bitter, Cooling, Galactagogue, Styptic, Tonic.

Medicinal Uses: Root – Dysentery.

406. *Typha elephantina Roxb.* (Typhaceae)

Syn.: *T. angustifolia* Watt. Non L.

Vern.: Aer.

Distribution: Common & gregarious in marshy & water logged places along roadsides and in ditches throughout the state.

Botany: Erect, rhizomatous herbs, with spongy linear leaves flat above and convex beneath. Spikes cylindrical, male flowers above, female below and up to 3 cm long gap in between. *Flowering & Fruiting:* Throughout the year.

Medicinal Parts: Whole plant.

Chemical Constituents: Seed yield a fatty oil.

Medicinal Properties: Whole plant – Cooling, Ahprodisiac.

Medicinal Uses: Whole plant – Burning sensation, Gonorrhoea, Inflammation, Leprosy, Splenic enlargement, Strangury.

407. *Uraria picta Desv.* (Fabaceae)

Syn.: *Hedysarum pictum* Jacq.

Vern.: Dabra.

Distribution: Growing as undergrowth of forest. Balaghat, Damoh, Hosahangabad, Mandla, Raisen, Sagar, Satna, Seoni, Shahdol, Sidhi.

Botany: An erect, little-branched perennial undershurbs; upper leaflets white-clouded; flowers purple; pods white or greyish, polished.

Flowering & Fruiting: August-January.

Medicinal Parts: Whole plant.

Chemical constituents: 5-hydroxy-7, 4-dimithoxy flavonol & 5 other flavonoids.

Medicinal Properties: Antibiotic, anti-inflammatory, alterative, carminative, stomachic, tonic.

Medicinal Uses: Bleeding piles, burning sensation, cough, diarrhoea, intermittent fever, rheumatism, ulcer.

408. *Urginea indica* (Roxb.) Kunth. (Liliaceae)

Syn.: *Drimia indica* (Roxb.) Jessop.

Vern.: Jangli kanda.

Distribution: In open dry situations, sometimes amongst stones and gravels in open places or shades. Rewa, Satna.

Botany: A small geophyte with ovoid, tunicated, white bulb, leaves radical, linear-lanceolate; flowers pale-brown, on a leafless scape, in few flowered racemes; capsule long, ellipsoid; seeds elliptic, black.

Flowering & Fruiting: March-May.

Medicinal Parts: Bulbs.

Chemical constituents: Bulb contain Glucosides, Scillian, Scillipicrin, Scillitin, Mucilage, Sugars, Calcium oxalate.

Medicinal Properties: Anthelmintic, cardio tonic, diphoretic, diuretic, expectorant, laxative.

Medicinal Uses: Asthma, blood pressure, cough, menorrhagia, paralysis, pneumonia fever, skin diseases, whooping cough.

409. *Urtica dioica* L. (Urticaceae)

Distribution: In moist habitats.

Botany: Erect herbs with stinging hairs; stems 4-angled stipules tree; leaves serrate; flowers dioecious.

Flowering & Fruiting: August-September.

Medicinal parts: Whole plant.

Chemical constituents: Vitamins, Carotenes, Acetylcholine, Histamine, 5-hydroxytryptamine, Lecithin.

Medicinal properties: The plant is Anti-rheumatic, Astringent, Anthelmintic, Antiasthmatic, Antidiarrhoeal, Antiscorbutic, Diuretic, Stimulant, Tonic.

Medicinal uses: The plant is useful in Menstrual disorders, Nephritic trouble, as Blood purifier; Haemorrhages from Kidney, Utereus & Nose; Vomiting of Blood, Sciatica.

410. *Ventilago maderaspatana Gaertn.* (Rhamnaceae)

Syn.: *V. calyculata* Tul.; *V. denticulata* Willd.

Vern.: Keonti.

Distribution: Balaghat, Damoh, Dhar, Hosahangabad, Indore, Jabalpur, Mandla, Narsimhpur, Sagar, Satna, Seoni, Shahdol, Sidhi.

Botany: Large woody climbers, with Greenish flowers & Yellowish globular nuts.

Flowering & Fruiting: September-April.

Medicinal Parts: Bark.

Chemical Constituents: Stem bark – Friedelin.

Root bark – Trihydroxy-methyl anthranolmonomethyl Ether, Emodinmonomethyl ether.

Fruits, Leaves, Stems – Lupeol, β-Sitosterol & Glucoside.

Medicinal Properties: Bark – Alexeteric, Astringent, Bitter, Carminative, Depurative, Digestive, Stimulant, Stomachic, Thermogenic, Tonic, Vulnerary.

Medicinal Uses: Bark – Colic, Dyspepsia, Erysipelas, Fever, Flatulence, General debility, Leprosy, Pruritus, Scabies, Skin diseases.

411. *Vernonia cinerea* (L.) Less (Asteraceae)

Syn.: *Conyza cinerea* L.

Vern.: Sahadevi.

Distribution: Very common throughout the State, on roadsides and in waste places.

Botany: Erect or decumbent herbs, with ovate to obovate or variable leaves. Heads violet or pink. Achenes terete.

Flowering & Fruiting: Throughout the year.

Medicinal Parts: Root, flower, seeds, whole plant.

Chemical constituents: Plant contains Lupeol acetate, alpha – amyrin, acetate, glucoside, Quercetin, β – sitosterol, stigma sterol, œ – spina sterol, potassium chhoride, œ – amyrin, lupeol.

Flower contains Luteolin, Iso – orientin, chrysoeriol.

Medicinal Properties: Plant – Astringent, cold, stomachic, sweet, tonic.

Medicinal Uses: Plant – Asthma, bronchitis.

412. *Vetiveria zizanioides* (L.) Nash. (Poaceae)

Syn.: *Andropogon muricatus* Retz.

Distribution: Balaghat, Damoh, Gwalior, Hosahangabad, Rewa, Sagar.

Botany: A densely tufted, erect perennial aromatic grasses; sessile spikelets awnless; yellow scabrous when mature.

Flowering & Fruiting: August-November.

Medicinal Parts: Roots.

Chemical constituents: Root contains – Essential oil, resin, acidic substance, iron, calcium, salts, citronellal, geraniol, terpene, Lignin.

Medicinal Properties:

Roots – Acrid, anthelmintic, aromatic, bitter, carminative, digestive, diuretic, expectorant, febrifuge, refrigerant, stimulant, stomachic, styptic, sweet.

Medicinal Uses: Roots – Anaemia, asthma, burning sensation, colic, dyspepsia, fever, hiccough, leprosy, menorrhagia, skin diseases, ulcers, vomiting.

413. *Vigna radiata* (L.) Wilczek (Fabaceae)

Syn.: *Phaseolus radiatus* L. non Roxb. & Auct.

Vern.: Moong.

Distribution: Cultivated. Bhopal, Hosahangabad, Indore, Mandla, Jabalpur, Raisen, Sagar, Shahdol, Shivpuri.

Botany: Twiners, with trifoliate leaves & thin, subcylindrical pods fruits. *Flowering & Fruiting:* September-December.

Medicinal Parts: Seeds.

Medicinal Properties: Seeds – Acrid, Aphrodisiac, Astringent, Constipating, Depurative, Digestive, Diuretic, Galactagogue, Styptic, Sweet.

Medicinal Uses: Seed – Asthma, Dropsy, Fever, General debility, Gonorrhoea, Leucoderma, Scabies, Skin diseases, Strangury.

414. Vigna radiata (L.) Wilczek Var. Sublobata (Roxb.) Verdc. (Fabaceae)

Syn.: *Phaseolus sublobatus* Roxb.

Vern.: Jungli Moong.

Distribution: Hosahangabad, Indore, Mandla, Shahdol.

Botany: Twiners, with trifoliate leaves & thin, subcylindrical pods fruits. *Flowering & Fruiting:* September-December.

Medicinal Parts: Roots.

Medicinal Properties: Roots – Anti-Inflammatory, Aphrodisiac, Bitter, Constipating, Depurative, Digestive, Emollient, Febrifuge, Refrigerant, Sweet.

Medicinal Uses: Roots – Burning sensation, Colic, Cough, Diarrrhoea, Dyspepsia, Faltulence, Fever, General debility, Haemorrhages, Inflammation, Leprosy, Seminal weakness, Skin diseases.

415. *Vigna unguiculata* (L.) Walp. ssp. *Cylindrica* (L.) Eselt. (Fabaceae)

Syn.: *V. sinensis* (L.) Hassk.; *V. catjang* (Burm. f.) Walp.

Vern.: **Lobia.**

Distribution: Cultivated.

Botany: A suberect or twining annual. Leaflets 3, entire or slightly lobed. Flowers Yellow or Reddish. Pods linear, subtorulose.

Flowering & Fruiting: September-November.

Medicinal Parts: Seeds.

Chemical Constituents: Seed contains – Carotene, Thiamine, Riboflavin, Niacin, Folic acid, Vitamin-C, Phytohemagglutinin, □-Cedrene, Hexylacetate, Limonene, Nonanal, β-Ocinene, □-Pinene, β-Pinene.

Medicinal Properties: Seeds – Anaphrodisiac, Anthelmintic, Appetiser, Astringent, Diuretic, Laxative, Liver tonic, Sweet.

Medicinal Uses: Seeds – Agalactia, Anorexia, Constipation, General debility, Helminthiasis, Jaundice, Strangury.

416. *Vitex negundo* L. (Verbenaceae)

Vern.: Negur.

Distribution: Common along roadsides, river bank and cultivated fields. Balaghat, Betul, Chhatarpur, Damoh, Hosahangabad, Indore, Jabalpur, Mandla, Panna, Sagar, Satna, Seoni, Shivpuri, Sidhi, Tikamgarh.

Botany: Evergreen shrubs, with 3 to 5-foliolate leaves. Flowers violet, in peduncled cymes. Fruits ovoid, 4-celled.

Flowering & Fruiting: June-December.

Medicinal Parts: Whole plant, leaves, fruits, seeds.

Chemical constituents: Volatile oil, Alkaloids (Nergundin & Hydrocotylon), Vitamin – C, Carotene, flavon, Tannic acid, Hydroxy benzoic acid, Resin, Mallic acid.

Medicinal Properties: Acrid, anthelmintic, astringent, bitter, cephalic, heating, Pungent, stomachic.

Medicinal Uses: Asthma, biliousness, bronchitis, enlargement of the spleen, eye diseases, inflammation, leucoderma.

Ayurvedic preperations: *Nirgundi kalpa, Nirgundi taila, Vishgarbha taila, Nirgundi ghrit, Nirgundi kwath.*

417. *Vitis vinifera Dunal.* (Vitaceae)

Vern.: *Angoor.*

Distribution: Often cultivated for its edible fruits.

Botany: A large, deciduous, climber with long bifid tendrils.

Flowering & Fruiting: February-May.

Medicinal Parts: Fruits, Leaves.

Chemical constituents: Fruits contain – Gum, tannin, grape-sugar, citric, racemic, mali acids; chlorides of potassium & sodium, sulphate of potash, tartarate of lime, mangnesia, alulm, iron.

Medicinal Properties: Fruits – Appitiser, cardiotonic, depurative, digestive, diuretic, emollient, expectorant, febrifuge, laxative, nervine tonic, nutritive, refrigerant, stomachic, sweet, tonic.

Medicinal Uses: Fruits – Anaemia, asthma, blood pressure, bronchitis, burning sensation, cough, fever, flatulence, general debility, jaundice, kidney diseases, leprosy, liver diseases, skin diseases, vertigo.

418 Withania somnifera (L.) Dunal (Solanaceae)

Vern.: Ashwagandha.

Distribution: In wastelands.

Botany: Hoary-tomentose undershrubs, with elliptic-lanceolate or obovate leaves, bisexual flowers and berries enclosed in the persistent calyx.

Flowering & Fruiting: August – December.

Medicinal Parts: Root, leaves, fruits, seeds, whole plant.

Chemical constituents: Root contains Alkaloids (Somniferon), Withaminon, Wasamin, Sugars, Glycosides, Amino acids, essential oils, Withaniol, Hentriacontane, Phytosterol & Oils.

Leaves contain Withanoloids alkaloids, glycosoids, glucose, amina acids.

Stem contains Proteins, amino acids, Tannin, flavonoids, Calcium, Iron.

Medicinal Properties: Tuber – Acrid, alterative, anthelmintic, aphrodisiac, bitter, tonic.

Medicinal Uses: Tuber – Asthma, bronchitis, inflammation, insomnia, psoriasis, scabies, ulcers.

Ayurvedic preperations: *Ashwagandhadi churna, Ashwagandha rasayan, Ashwagandha ghrit, Ashwagandharishta, Narayan taila, Ashwagandhavaleha, Maharasnadi yoga, Sukumaraghrit, Saubhagyasunthipaka.*

419. *Woodfordia fruticosa* (L.) Kurz. (Lythraceae)

Syn.: *W. floribunda* Salisb.

Vern.: Dhao.

Distribution: Almost throughout the State.

Botany: Shrubs, with broadly lanceolate leaves and deep-red flowers in axillary cymes. Capsules oblong.

Flowering & Fruiting: January-May.

Medicinal Parts: Bark, Flower.

Chemical constituents: Leaves contain Ellagic acid & polyphenols, lawsone & tannins.

Stem contain Octacosnol, β-sitosterol.

Flowers contain Hecogenin & gallic acid, tannins, soluble nontans & sugars.

Medicinal Properties: Bark – Acrid, alexiteric, anthelmintic, cooling, Pungent, tonic.

Medicinal Uses: Bark – Blood diseases, dyspepsia, erysipelas, leprosy.

Flowers – Leucorrhoea, menorrhagia, toothache.

Ayurvedic preperations: *Dhatakayadi churna, Dhatakyadi taila, Kutajarista, Laghugangadhar churna, Abhyarista, Pippalayasav, Kanakasava, Khadirarista, Parthadyarista, Ashokarista.*

420. Wrightia tinctoria (Roxb.) R. Br. (Apocynaceae)

Syn.: *W. rothii* G. Don.; *Nerium tinctorium* Roxb.

Vern.: Indrajau.

Distribution: Rare. Hoshangabad.

Botany: A small tree with glabrous young parts; leaves elliptic-oblong; flowers in lax, terminal corymbose, dichotomous cymes. Fruits follicles. Seeds with a tuft of hairs at the base.

Flowering & Fruiting: March-October.

Medicinal Parts: Root, Leaves, Bark, Seed.

Chemical constituents: Seed contains – Fixed oil.

Bark – β-sitosterol, β-amyrin, lupeol benzoate.

Medicinal Properties: Bark & Seeds – Acrid, anthelmintic, antipyretic, aphrodisiac, astringent, bitter, carminative, digestive, stomachic, thermogenic.

Leaves – Acrid, anodyne, thermogenic.

Medicinal Uses: Bark & Seeds – Burning sensation, colic, diarhroea, dropsy, dysentery, fever, flatulence, helminthiasis, leprosy, psoriasis.

Leaves – Hypertension.

421. *Xanthium strumarium* L. (Asteraceae)

Syn.: *X. indicum* Koenig ex Roxb.

Vern.: Adhasisi.

Distribution: Throghout the State as an Weed.

Botany: Foetid-smelling unarmed herbs, about 1m high. Leaves ovate-suborbicular. Heads greenish yellow, in terminal and axillary racemes.

Flowering & Fruiting: January – February.

Medicinal Parts: Whole plant.

Chemical constituents: Seed contains Xanthostrumarin acid, oxalic acid.

Fruits contain Strumaroside, stigmasterol, xanathin, xanthinin, xanthumin, xanthanol.

Medicinal Properties: Acrid, alexiteric, alterative, anthelmintic, antipyretic, appetiser, cooling, digestive, fattening, laxative, tonic.

Medicinal Uses: biliousness, epilepsy, fever, leucoderma, salivation.

422. *Zingiber officinale Rosc.* (Zingiberaceae)

Vern.: Adrak.

Distribution: Cultivated. Bhopal, Chhatarpur, Indore.

Botany: A small herb with horizontal, fleshy rhizome; leaves distichous, narrow, oblong-lanceolate; Flowers on long spike and are of purple colour on spike, small.

Flowering & Fruiting: September-October.

Medicinal Parts: Rhizome.

Chemical constituents: Rhizome contains Essential oil, Fibres, Proteins, Starch, Carbohydrates, Resin, Glutamin, Thrionin, Free amino acids, Zingiberol, Zingiberin, Glutamic acid, Aspartic acid.

Medicinal Properties: Rhizome – Analgesic, aphrodisiac, appitizer, aromatic, blood purifier, carminative, digestive, expectorant, febrifuge stimulant, thermogenic.

Medicinal Uses: Rhizome – Arthritis, asthma, chronic bronchitis, cold, cough, dysuria, flatulence, gout, joint problems, lumbago, muscular trouble, rheumatism, sciatica, slip disc.

Ayurvedic preperations: *Amritarishta, Adrakkhand, Panchsam churna, Samsharkar churna, Rasnadi kwath, Saubhagyasunthi paka, Shunti sura, Sunthi panak, Talishadi churna, Lavangadi churna, Hingvastak churna, Trikatu churna, Laghusutasekhara and Vyosadhi ghrit.*

423. *Ziziphus mauritiana Lam.* (Rhamnaceae)

Syn.: *Z. jujuba* Mill.; *Z. vulgaris* Lam.; *Z. sativa* Gaertn.

Vern.: Ber.

Distribution: Balaghat, Damoh, Hosahangabad, Indore, Mandla, Panna, Sagar, Satna, Shivpuri, Sidhi.

Botany: Evergreen small trees or shrubs, with droopling branches armed with prickles. Leaves ovate-elliptic to suborbicular, tomentose beneath. Flowers greenish – yellow, in axillary cllusters. Drupes fleshy.

Flowering & Fruiting: September-June.

Medicinal Parts: Bark, leaves, fruits.

Chemical constituents: Fruits contains Sugars & mucilage.

Medicinal Properties: Root – Bitter, cooling.

Leaf – Antipyretic, bitter, cooling.

Fruits – Aphrodisiac, cooling, indigestible, laxative, tonic.

Seed – Aphrodisiac, acrid, sweet.

Medicinal Uses: Root – Biliousness, headache.

Bark – Boils, diarrhoea, dysentery.

Leaf – Biliousness, diarrhoea, obesity.

Fruits – Biliousness, blood diseases, vomiting.

Seed – Asthma, burning sensation, cough, eye diseases, leucorrhoea, vomiting.

424. *Ziziphus nummularia* (Burm. F.)Wt. & Arn. (Rhamnaceae)

Syn.: *Z. rotundifolia Lam.; Rhamnus nummularia Burm f.*

Vern.: Jharberi.

Distribution: Hosahangabad, Shivpuri, Sidhi.

Botany: Much-branched, prickly shrubs, with zigzag branches & orbicular-ovate leaves tomentose above. Flowers greenish-yellow, in axillary cymes. Drupes globose, red when ripe.

Flowering & Fruiting: September-October.

Medicinal Parts: Leaves, Fruits.

Chemical Constituents: Twig bark contains Tannin.

Root bark contains Cyclopeptide alkaloids Nummularine A, B, C; Mucronine-D & Amphibine-H.

Medicinal Properties: Fruits – Appetiser, Astringent, Cooling, Sour, Stomachic, Sweet.

Medicinal Uses: Leaves – Cold, coughs, Scabies, Skin diseases.

Fruits – Bilious affections.

425. *Ziziphus oenoplia* (L.) Mill. (Rhamnaceae)

Vern.: Makoya.

Distribution: Balaghat, Bhopal, Damoh, Gwalior, Hosahangabad, Panna, Raisen, Rewa, Sagar, Satna, Seoni, Shahdol, Shivpuri, Sidhi.

Botany: A straggling or climbing shrub. Prickles paired one straight the other recurved. Leaves ovate, pointed; base oblique. Flowers in short axillary, dichotomous cymes. Drupes globose or obovoid, Black, shining. *Flowering & Fruiting:* May-October.

Medicinal Parts: Roots.

Chemical Constituents: Bark contain tannin.

Stem & Root bark contain Cyclopeptide alkaloids Zizyphine-A & Zizyphine-B.

Medicinal Properties: Roots – Anthelmintic, Antiseptic, Astringent, Bitter, Digestive.

Medicinal Uses: Roots – Ascaris Infection, Hyperacidity, Stomachalgia, Wounds.

Glossary of Medical Terms

Abortifacient	an agent that induces abortion
Acrid	biting, pungent
Adenitis	inflammation of a gland or group of glands (or glandlike structures)
Agalactia	absence or failure of secretion of milk
Alexipharmic	antidote to poison
Alexiteric	protective to infectious diseases
Alopecia	loss of hair: a malady in which the hair falls from one or more circumscribed round or oval areas, leaving the skin smooth and white
Alterative	causing a favourable change in the disordered functions of the body or metabolism
Amenorrhoea	failure of menstruation
Amentia	an arrest of the development of the mind from birth to early age
Anaemia	lack of enough blood causing paleness
Analgesic	an anodyne
Anaphrodisiac	having the power to lessen or inhibit sexual feeling
Anodyne	a medicine that allays pain
Anorexia	a condition of having lost the appetite for food
Anthelmintic	destroying or expelling worms
Antibacterial	active against bacteria
Antidote	an agent which neutralises or opposes the action of a poison
Anti-inflammatory	acting against inflammation
Anti-leprotic	any agent which prevents or cures leprosy
Antilithic	an agent which prevents the formation of calculi
Anti-periodic	preventing the regular recurrence of a disease
Antipyretic	counteracting fever
Antiscorbutic	acting against scurvy
Anti-septic	a chemical sterilising substance to kill or control pathogenic microbes

Antispasmodic	opposing Spasms
Anuria	failure of the kidneys to produce urine
Aperient	A laxative
Aphrodisiac	drug which stimulates sexual desire
Areata	ocurring in circumscribed areas or patches
Arthritis	inflammation of joints
Ascities	an abnormal accumulation of a fluid in the peritoneal cavity
Astringent	a drug which contracts the organic tissues & canals of the body, & therby checking or deminishing bleeding of excessive discharges
Azoospermia	the comlete absence of sperm from the seminal fluid
Baldness	absence of hair from areas where it normally grows
Beriberi	a deficiency disease caused by imbalance of carbohydrate and vitamin B
Biliousness	a symptom complex comprising nausea
Boil	tender inflamed area of the skin containing pus
Bronchitis	inflammation of the bronchi
Bronchodilator	an agent that causes widening of the air passages by relaxing bronchial smooth muscle
Bubo	an inflammatory swelling of a lymph gland
Calculi	a hard pebble-like mass formed within the body, particularly in the gall bladder or anywhere in the urinary tract
Cancer	the abnormal and uncontrolled division of cells that then invade and destroy the surrounding tissues
Carbuncle	an infection of the skin and subcutaneous tissue by Staphylococcus aureus
Carminative	drug curing flatulence
Catarrh	inflammation of a mucous membrane
Cathartic	having the power of cleansing the bowels-purgative
Cephalagia	headache
Cephalic	a remedy for disorders of the head
Cholera	a severe infectious epidemic disease due to Vibrio cholerae
Chronic bronchitis	the patient coughs up excessive mucus secreted by enlarged bronchial mucous glands

Conjunctivitis inflammation of conjunctiva
Colic a severe spasmodic griping pain
Consumption pulmonary tuberculosis
Contraceptive drug which prevents unwanted pregnancy
Convulsion a violent involuntry contraction of the skeletal musculature
Cystitis inflammation of a bladder, especially the urinary bladder
Dandruff dead scarf-skin seperating in small scales and entangled in the hair
Deafness partial or total loss of hearing in one or both ears
Demulcent soothing
Depurative an agent that purifies the blood
Diabetes here it refers to diabetes mellitus: a disorder of carbohydrate metabolism in which sugars in the body are not oxidized to produce energy due to lack of the pancreatic hormone-insulin. The accumulation of sugar leads to its appearance in the blood
Diaphoretic a drug which induces perspiration
Diarrhoea frequent bowel evacuation or the passage of abnormally soft or liquid faeces
Diptheria a specific infectious disease caused by virulent strains of a bacillus
Diuretic promoting the discharge of urine
Dropsy an excessive accumulation of a clear or watery fluid in any of the tissues or cavities of the body
Dysentery an infection of the intestinal tract causing severe diarrhoea with blood and mucus
Dysmenorrhoea painful menstruation
Dysopia defective vision
Dyspepsia indigestion
Dysuria difficulty or pain while passing urine
Earache pain in the ear
Eczema a noncontagious inflammatory disease of the skin with much itching and burning
Elephantiasis gross lymphatic oedema of the limbs leading to hypertrophy
Emetic causing vomiting

Emollient	softening
Epilepsy	an affection of the nervous system resulting from excessive or disordered discharge of cerebral neurons
Eruptions	the outbreak of a rash
Erysipelsa	an inflammatory disease generally on the face marked by a bright redness of the skin
Expectorant	aiding the secretion of the mucous membrane of the air passages & the removal of fluid by spitting
Fatigue	mental or physical tiredness, following prolonged or intense activity
Febrifuge	that which drives off fever
Fever	a rise in body temperature above the normal
Fistula	an open channel from the anus or rectum to the skin near the anus
Flatulence	distention of stomach
Foetid	bad-smelling, fetid
Gleet	chronic discharge of thin mucus from the vagina
Goitre	enlargement of the thyroid gland
Gonorrhoea	an inflammatory disease of the urinogenital passagaes characterised by pain and discharge
Gout	a disease of purine metabolism characterised by attacks of arthritis with an associated raised level of serum uric acid
Haematemesis	vomitting of blood
Haematinic	any substance which is required for the production of RBCs & its constituents
Haemoptysis	spitting of blood
Haemorrhage	the escape of blood from a ruptured blood vessel, externally or internally
Haemorrhagic diseases	associated with or resulting from blood loss
Haemostatic	styptic
Hallucination	a false perception of something that is not really there
Headache	pain felt deep within the skull
Helminthiasis	morbid state due to infestation with worms
Helminthic	Anthelmintic, vermifugal
Hemicrania	headache confined to one side

Hepatomegaly	enlargement of the liver
Hiccough	abrupt involuntary lowering of the diaphragm and closure of the sound-producing folds at the upper end of the trachea, producing a characteristic sound as the breath is drawn in
Hydrocele	a circumscribed collection of fluid in the tunica vaginalis testis
Hydrophobia	exaggerated fear of water as in rabies
Hypertension	high arterial blood pressure
Hysteria	a neurotic disorder with varying symptoms
Impotency	inability to perform the sexual act due to failure of the reflex mechanism
Inflammation	the body's response to injury, which may be acute or chronic
Insecticidal	any agent which kills or destroys insects
Insomnia	the condition a being unable to sleep/
Intermittient fever	a fever that rises, subsides, then returns again
Itcing	discomfort or irritation of the skin, prompting the sufferer to scratch or rub the affected area
Jaundice	a yellowing of the skin or whites of the eyes, indicating excess bilirubin(a bile pigment) in the blood
Laxative	a drug used to stimulate or increase the frequency of bowel evacuation, or to encourage the passage of a softer or bulkier stool
Leprosy	a chronic disease, caused by the bacterium Mycobacterium leprae, that affects the skin, mucous membranes, and nerves
Leucoderma	loss of pigments in areas of the skin, resulting in the appearance of white patches or bands
Leucoma	white opacity occurring in the cornea
Leucorrhoea	whitish or yellowish discharge of mucus from the vaginal opening
Lumbago	pain in lower back
Mammillitis	inflammation of the nipple
Menorrhagia	excessive or prolonged menstruation
Migrain	condition resulting from spasm and subsequent overdilatation of certain arteries in the brain, which causes a recurrent throbbing headache that characteristically affects one side of the head

Miscarriages	the expulsion or removal of an embryo or fetus from the uterus at a stage of pregnancy when it is incapable of independent survival
Mumps	an acute infectious disease caused by a virus
Nasitis	inflammation of the nose
Nightblindness	the inability to see in dim light or at night
Obesity	an excessive accumulation of fat in the body
Oedema	excessive accumulation of fluid in the body tissues
Oleaginous	Oily
Ophthalmia	a term usually applied to conjunctivitis
Ophthalmopathy	any disease of the eye
Otalgia	pain in the ear
Otorrhoea	any discharge from the ear, commonly a purulent discharge in chronic middle ear infection
Paralysis	a muscle weakness that varies in its extent, its severity, and the degree of spasticity or flaccidity according to the nature of the underlying disease and its distribution in the brain, spinal cord, peripheral nerves, or muscles
Pectoral diseases	effective in diseases of the chest
Peptic ulcer	a breach in the lining(mucosa) of the digestive tract produced by digestion of the mucosa by pepsin and acid
Piles	enlargement of the normal spongy blood-filled cushions in the wall of the anus, usually a consequence of prolonged constipation or ocassionally, diarrhoea
Pneumonia	a general disease in which the essential lesion is an inflammation of the spongy tissue of the lung with consolidation of the alveolar exudate
Pneumonosis	any lung disorder
Polyuria	the production of large volumes of urine, which is dilute and of a pale colour
Procreant	the drug which begets
Prophylactic	pertaining to the prevention of the development of a disease
Prurigo	an eruption of the skin causing severe itching
Pruritus	itching

Pyorrhoea	disease of the tissues that support and attach the teeth-the gums, periodontal membrane, and alveolar bone
Refrigerant	cooling
Remittent fever	a remittent fever is one in which body temperature flactuates but does not return to normal
Renal calculi	calculi relating to kidney
Resolvent	causing resolution of a tumour or swelling
Retention of urine	inability to pass urine, which is retained in the bladder
Revulsive	causing revulsion in drawing away of blood from a pathological area to another area
Rheumatalgia	rheumatic pain
Rheumatism	any disorder in which aches and pains affect the musclas and joints
Rheumatoid arthritis	it is the second most common form of arthritis. It typically involves the joints of the fingers, wrists, feet and ankles with later involvement of the hips, keens, shoulders and neck
Rhinitis	inflammation of the nasal mucous membrane
Ringworm	fungus infection of the skin, the scalp, or the nails
Rubefacient	reddening of the skin
Scabies	sarcoptic infestation of the human skin particularly a contagious skin disease caused by invasion of the epidermis
Sciatica	pain felt down the back and outer side of the thigh, leg, and foot
Scrofula-	tuberculous cervical adenitis with or without ulceration
Scurvy	a deficiency disease due to lack of vitamin C
Sinusitis	inflammation of one or more of the mucous-membrane-lined air spaces in the facial bones that communicate with the nose
Slipdisc	a colloquial term for a prolapsed intervertebral disc
Small Pox	an acute infectious virus disease causing high fever and a rash that scars the skin
Somatalgia	pain in the body

Soporific	inducing sleep
Sore throat	pain at the back of the mouth, commonly due to bacterial or viral infection of the tonsils
Spasmolytic	a drug that relieves spasm of smooth muscle
Splenopathy	any disease of the spleen
Spermatorrhoea	the involuntary discharge of semen without orgasm
Stomachache	pain in the stomach
Stomachic	a medicine that increases the functional activity of the stomach
Stomatitis	generalised inflammation of the oral mucosa
Strangury	severe pain in the urethra referred from the base of the bladder and associated with an intense desire to pass urine
Styptic	having the power to arrest bleeding
Sudorific	an agent which induces sweating
Sun stroke	raised body temperature (pyrexia), absence of sweating, and eventual loss of consciousness due to failure or exhaustion of the temperature-regulating mechanism of the body
Syphilis	contagious venereal disease
Tetanus	an infective disease due to toxin of Clostridium tetani
Tuberculosis	an infectious disease caused by the bacillus Mycobacterium tuberculosis and characterized by the formation of nodular lesions in the tissues
Tumour	any abnormal swelling in or on a part of the body
Typhoid	an infection of the digestive system by the bacterium Salmonella typhi, causing general weakness, high fever, a rash of red spots on the chest and abdomen
Ulcers	a break in the skin extending to all its layers, or a break in the mucous membrane lining the alimentary tract, that fails to heal and is often accompained by inflammation
Urethrorrhoea	abnormal discharges from the urethra
Urticaria	an itchy rash resulting from the release of histamine by mast cells
Uteralgia	pain in the uterus

Veneral diseases	any disease transmitted by sexual intercourse
Verminosis	infestation with vermin
Vertigo	Dizziness
Vesical calculi	stone in the urinary bladder
Vomiting	the reflex action of ejecting the contents of the stomach through the mouth
Vulnerary	an agent that promotes healing of wounds
Warts	a circumscribed cutaneous excrescence
Whooping cough	an acute infectious disease, primarily of infants, caused by the bacterium Bordetella pertusis, which affects the respiratory tract
Wounds	a break in the structure of an organ or tissue caused by an external agent

Bibliography

Ambasta, S. P. 1986 (Ed.) *The useful plants of India. CSIR,* New Delhi.

Anonymous. 1948-1976. *The Wealth of India. CSIR,* New Delhi.

Bhandari, C. 1938. *Vanaushdhi chandrodaya* 1-10. (Hindi edition). Gyan Mandir, Indore.

Chopra, R.N. Nayar, S.L. and Chopra, I.C. 1956. *Glossary of Indian medicinal plants.* CSIR.

Chopra, R.N. Badhwar, R.L. and Ghosh, S. 1965. *Poisonous plants of India.* Vol. I & II. ICAR, New Delhi.

Gogte, V.V.M. Vaidya. 2000. *Ayurvedic pharmacology and Therapeutic uses of medicinal plants* (Dravyaguna Vignyan). Bhartiya Vidya Bhawan, Mumbai.

Gupta, A.K. & Tandon, N. 2004(Edited). Reviews on Indian Medicinal Plants. *Vol.* I, II, III & IV. *Indian Council of Medical Research. New Delhi.*

Handa, S.S. & Kapoor, V.K. 1999. *Pharmacognosy.* Vallabh Prakashan, Delhi.

Jadhav, D. 2008. "Medicinal Plants of India: A Guide to Ayurvedic & Ethnomedicinal uses of Plants." Vol.-I. *Scientific Publishers,* Jodhpur.

Jadhav, D. 2008. "Medicinal Plants of India: A Guide to Ayurvedic & Ethnomedicinal uses of Plants." Vol.-II. *Scientific Publishers,* Jodhpur.

Jadhav, D. 2008. "Medicinal Plants of Madhya Pradesh & Chhattisgarh." *Daya Publishing House,* New Delhi.

Jadhav, D. 2009. "Medicinal Plants of India: A Guide to Ayurvedic & Ethnomedicinal uses of Plants." Vol.-III. *Scientific Publishers,* Jodhpur.

Jain, S. K. 1999. *Medicinal plants.* N.B.T. New Delhi.

Jain, S. K. 2002. *Vanaspati Sabdakosh: Upyogi paudhon ka Hindi-Latin-Angreji Sabdakosh.* (Hindi edition). Scientific publishers, Jodhpur.

Joshi, S. G. 2007. *Medicinal Plants.* Oxford & IBH Publishing Co. Pvt. Ltd., New Delhi.

Khanna, K. K., Anand Kumar, R. D. Dixit & N. P. Singh. 2001. Supplement to the Flora of Madhya Pradesh. BSI. Calcutta.

Parik, G. 2006-06 (Edited). *Vanausdhi Ratanakar.* 1-9. (Reprint). (Hindi edition). Vijaygarh, Aligarh.

Khory, J.K. 1981. *Materia medica of India & their Therapeutics,* Neeraj Publishing house, New Delhi.

Kirtikar, K.R. & Basu, B.D. 2004. *Indian medicinal plants,* Vol. I, II, III & IV (Reprint). Bishen Singh Mahendra Pal Singh, Dehradun.

Krishnanand, Swami 2004. *Gaon me Aushad ratna* (Hindi edition). Part 1-3. Krishna Gopal Ayurveda Bhawan (Dharmarth Trust), Kaleda, Rajasthan.

Lindley, John. 1981. *Flora medica, Ajay book service*, New Delhi.

Mishra, B. & Vaisya, R. 2004. B*havaprakasa*, Vol. I & II. Chaukhambha Sanskrit Sansthan, Varanasi.

Mudgal, V. Khanna, K.K. & Hajra, P.K. 1997. *Flora of Madhya Pradesh*. Vol. II. BSI. Calcutta.

Mukerji, B. 1953. *The Indian pharmaceutical Codex*, New Delhi.

Namjoshi, A.N. 1960. *Studies in Pharmacognosy of Ayurvedic Drugs*. Board of Research on Ayurveda, Bombay.

Oommachan, M. 1977. *The flora of Bhopal.* J. K. Jain brothers, Bhopal.

Pandey, G.S. 1969. *Bhavaprakash Nighantu* (With Commentary by Dr. K.C. Chunekar). Chowkhamba Sanskrit Sansthan, Varanasi.

Panigrahi, A.K. & Alka, S. 2000. *Glossary of useful and economically important plants*. New Central Book agency. Calcutta.

Paranjpe, P. *Indian medicinal plants: forgotten healers*. 2005. Chaukhamba Sanskrit Pratisthan, Delhi.

Printya, J.V. 1963. *The Sushruta Samhita*. Vol. I&II, Motilal Banarsidas. Delhi.

Rastogi & Mehrotra 1997. *Compendium of Indian medicine Plant*, Vol. I, II, III, IV & V. CDRI, Lucknow and PID Delhi.

Verma, D. M. *Balakrishan,* N. P. & Dixit, R. P. 1994. Flora of Madhya Pradesh. Vol. I. BSI, Calcatta.

Sharma, R. K. and Bhagvan dash. 1976. *Charaka samhita*, Vol. I & II, Chaukhamba Sanskrit Series, Varansi, India.

Sharma, P.V. 2003. *Dravyaguna – Vijnana*. Chaukhamba bharti academy Varanasi.

Singh, N. P., K. K. Khanna, V. Mudgal & R. D. Dixit. 2001. *Flora of Madhya Pradesh*. Vol. III. BSI. Calcutta.

Warrier, P. K., V. P. K. Nambiar & C. Ramankutty.1994. *Indian medicinal plants*. Vol. I – V. oriental longum, Chennai.

Index of Local, English, Hindi & Sanskrit Names of Plants

SN	Botanical name	Local name	English name	Hindi name	Sanskrit name
1	*Abelmoschus esculentus*	Bhindi.	Lady's finger	Bhindi	Dindisa
2	*Abelmoschus moschatus*	Kasturi Bhindi	Vegetable musk	Musk-dana	Lata-kasturika
3	*Abrus precatorius*	Ratti	Crab's eye	Ratti	Aruna
4	*Abutilon indicum*	Kanghi	Indian mallow	Kanghi	Atibala
5	*Acacia arabica*	Babool	Indian gum arabic tree	Babul	Babbula
6	*Acacia catechu*	Khair	Cutch tree	Khair	Khadira
7	*Acacia farnesiana*	Bilayati Babool	Cassie flower	Vilayati kikar	Irimeda
8	*Acacia leucophloea*	Safed kikar	White babool	Safed babool	Arimedah
9	*Acacia pennata*	Chundria	–	Aila	Ari
10	*Acacia sinuata*	Shikakai	Soapnut acacia	Satala, Koci, Ritha	Saptala
11	*Acalypha indica*	Kuppi	Indian acalypha	Kuppi	Rudra
12	*Achyranthes aspera*	Apamarg	The prickly chaff	Latjira	Apamarg
13	*Adansonia digitata*	Gorakh-imli	Monkey bread tree	Gorakh-imli	Goraksi
14	*Adenanthera pavonina*	Rakta-kanchan	Coralwood tree	Baragunci	Tamrakah
15	*Adhatoda zeylanica*	Adusa	Vasak	Adusa	Vasaka
16	*Aegle marmelos*	Bilvapatra	Wood apple	Bel	Bilva
17	*Aerva lanata*	Gorakhganja	–	Gorakh ganja	Bhadra
18	*Agave Americana*	Cactus	Centary plant	Kamal cactus	Kantala
19	*Ageratum conyzoides*	Osdi	Goat weed	Sadoradi	Visamusti

20	*Ailanthus excelsa*	Maharukh	The tree of heaven	Maharuk	Mahanimba
21	*Alangium salvifolium*	Ankol	Sageleaved alangium	Ankol	Ankota
22	*Albizia lebbeck*	Sirus.	East Indian walnut	Siris	Sirisa
23	*Albizia odoratissima*	Chichwa.	Black siris	Kala siris	Bhusirisah
24	*Albizia procera*	Safed Siris	White Sirish	Safed Siris	Kinihi
25	*Allium cepa*	Pyaz	Onion	Piyaz	Palandu
26	*Allium sativum*	Lahsun	Garlic	Lasun	Rasona
27	*Aloe vera*	Gwarpatha.	Indian aloe	Gwarpa tha	Kumari
28	*Alstonia scholaris*	Saptparni.	Devil tree	Chatium	Saptaparna
29	*Alternanthera sessilis*	Guroo.	–	Saranti	Matsyaksi
30	*Amaranthus spinosus*	Kantewali chauli	Prickly amaranthus	Kantew ali chauli	Tanduliya
31	*Amaranthus tricolor*	Lal Saag	Chinese spinach	Chaulai	Marisa
32	*Ammania baccifera*	–	Blistering ammania	Dadama ri	Kurandika
33	*Amorphophallus campanulatus*	Jangali suran	Elephant foot yam	Jangali suran	Surana
34	*Ampelocissus latifolia*	Panibel	Jungle grape vine	Panibel	–
35	*Anacardium occidentale*	Kaju	The cashewnut	Kaju	kajutaka
36	*Andrographis paniculata*	Kalmegh	Creat	Kalmegh	Bhunimba
37	*Anethum graveolens*	Sowa	Dill	Soya	Satapuspa
38	*Anisomeles indica*	Gopali	Malabar catmint	Kala bhangra	–
39	*Annona reticulata*	Ramphal	Bullock's heart	Raamph al	Ramaphala
40	*Annona squamosa*	Seetaphaal	Custard apple	Seetaphaal	Seetaphala
41	*Anogeissus latifolia*	Dhawra	Button tree	Dhawra	Dhava
42	*Anthocephalus cadamba*	Kadam	Kadam	Kadamb	Kadambah
43	*Arachis hypogaea*	Mung-phali	Ground nut	Mung phali	Mandapi
44	*Areca catechu*	Supari	Betal palm	Supari	Pugah
45	*Aristolachia bracteolata*	Kidamar	Worm killer	Kirmar	Kitamari
46	*Aristolochia indica*	Isari mul	Indian birthwort	Isarmul	Nakuli
47	*Argemone mexicana*	Satyanashi	Mexican poppy	Satyanas hi	Svarnaksiri
48	*Argyreia speciosa*	Samander ka pat	Elephant creeper	Samudra kapat	Samudra sosh

49	*Artemisia nilagirica*	–	Indian wormwood	Nagdona, Dauna	Damanah
50	*Artocarpus heterophyllus*	Kathal	Jackfruit tree	Kathal	Panasah
51	*Arundo donax*	–	Spanish cane	Bara nal	Nala
52	*Asclepias curassavica*	–	Wild epicacunha	Kauathori	Bharadvaji
53	*Asparagus racemosus*	Shataveri	Asparagus	Shatawar	Shatavari
54	*Averrhoa carambola*	–	Carambola apple	Kamarakh	Karmarangah
55	*Azadirachta indica*	Neem	Margosa tree	Neem	Nimbaka
56	*Bacopa monnieri*	Brahmi	Thyme-leaved gratiola	Brahmi	Brahmi
57	*Balanites aegyptiaca*	Hingot	Desert date	Hingot	Inguda
58	*Baliospermum montanum*	–	–	Danti	Danti
59	*Bambusa arundinacea*	Bans	Thorny bamboo	Bans	Vamsah
60	*Barleria prionitis*	Jhinti	Katsareya	Vajradanti	Saireyaka
61	*Barringtonia acutangula*	–	Small Indian oak	Samudra phal	Samudraphala
62	*Basella alba* L. var. *rubra*	Poi	Indian spinach	Poi	Upodika
63	*Bauhinia purpurea*	Kaliar	Butterfly tree	Kaliar	Kovidar
64	*Bauhinia racemosa*	–	Downy mountain ebony	Jhinjhora	Ashmantak
65	*Bauhinia variegata*	Kachnar	Camel's foot	Kachnar	Kanchanara
66	*Benincasa hispida.*	Gol kaddu	White gourd melon	Petha	Kusmandah
67	*Biophytum sensitivum*	Lajalu	Sensitive wood sorrel	Lajalu	Almbusa
68	*Blumea lacera*	–	Blumea	Kukarau nda	Kukundarah
69	*Boerhavia diffusa*	Punarnava	Horse puruni spreading	Gadhapur na	Punarnava
70	*Bombax ceiba*	Semal	Silk cotton	Semul	Salmali
71	*Borassus flabellifer*	–	Palmyra palm	Tar	Talah
72	*Boswellia serrata*	Salai	Indian olibanum tree	Salai	Sallaki
73	*Brassica juncea*	Rai	Indian mustard	Rayi	Sarsapah

74	*Brassica nigra*	Banarasi Rai	Black mustard	Rai	Asuri
75	*Bredelia retusa*	–	Spinous Kino tree	Ketakela	Asana
76	*Buchanania lanzan*	Chironji	–	Ciraumci	Priyalah
77	*Butea monosperma*	Palash	Flame of the forest	Palash	Palasha
78	*Caesalpinia bonduc*	Kantak karanj	Fever nut	Kantaka ranj	Kantaki karanja
79	*Cajanus cajan*	Arahar	Pigeon pea	Tuvar	Adhaki
80	*Calotropis gigantea*	Safed aak	Giant milk weed	Madar	Alarka
81	*Calotropis procera*	Aak	Milk weed	Akda	Arka
82	*Canavalia gladiata*	Bansem.	Borad bean	Lalkudu sumpal	Asisimbi
83	*Cannabis sativa*	Bhang	True hemp	Bhang	Bhanga
84	*Canscora decussata*	Sankhaphuli	Canscora	Samkhap huli	Sankhini
85	*Capparis decidua*	Karer	Caper plant	Karil	Karirah
86	*Capparis sepiaria*	–	Hedge caper	Kantari	–
87	*Capparis spinosa*	–	Flinders rose	Himsra	Cabra
88	*Capparis zeylanica*	Ardanda	Ceylon caper	Ardanda	Karambha
89	*Capsicum annuum*	Lal mirch	Chillis	Lalmircah	Katuvirah
90	*Capsicum frutescens*	Mirch	Wild chilli pepper	Mirch	–
91	*Cardiospermum halicacabum*	Kanphuta	Ballom vine	Kanphuti	Sakralata
92	*Careya arborea*	–	Wild guava	Kumbi	Kumbhi
93	*Carica papaya*	Papit	Papaya	Pappita	Erandakakati
94	*Carissa carandas*	Karonda	Christ's thorn	Karaun da	Avighnah
95	*Caryota urens*	Mari	Fish tail palm	Mari	Sritalah
96	*Cassia fistula*	Amaltas	Indian laburnum	Amaltas	Chaturangulo
97	*Cassia sophera*	Kasaunda	*Cassia sophera*	Kasaundi	–
98	*Catharanthus roseus*	Sadabahar	Periwinkle	Sadabah ar	Nityakalyani
99	*Cayratia carnosa*	Amal-bel	Fox grape	Amalbel	Gandirah
100	*Celastrus paniculatus*	Malkangani	Climbing staff plant	Malkang ani	Jyotismati
101	*Celosia argentea*	Safed murga	Quail grass	Safed murga	Vitunna
102	*Centella asiatica*	Mandukaparni	Indian pennywort	Brahma manduki	Mandukaparni

103	*Centipeda minima*	Nakchhikni	Spreading Sneeze Weed	Nakchhi kni	Chhikka
104	*Chenopodium album*	Bathu	White goosefoot	Bathua sag	Vastukah
105	*Chlorophytum tuberosum*	Safed musli	White musali	Safed musli	Mushli
106.	*Chrozophora rottleri*	–	Turnsole	–	Suryavarta
107.	Cicca distichia	Harfarebari	Country gooseberry	Chalmeri	Lavali
108.	*Cicer arietinum*	Chana	Bengal gram	Cana	Canakah
109.	*Cinnamomum tamala*	Tejpat	Indian cassia lignea	Tejpatt	Tamalapattra m
110.	*Cissampelos pareira*	Akauadi	False pareira	Karjeuri	Ambashtha, patha
111.	*Citrus aurantium*	Moosambi	Sweet Orange	Malta	Nagaranga
112.	*Citrus limon*	–	Lime	Jamiri nimbu	Jambirah
113.	*Citrus maxima*	Chakotara	Chinese grapefruit	batawi-nimbu	Karuna
114.	*Citrus medica*	Bara nimbu	Citron	Baranim bu	Matulungha
115.	*Clematis triloba*	Murhari	–	Churaha r	Morata
116.	*Cleome gynandra*	Hulhul	Spider flower	Hulhul	Suvarchala
117.	*Cleome viscosa*	Hulhur	Wild mustard	Hurhur	Ajagandha
118.	*Clerodendrum inerme*	Choti Arni.	–	Lanjai	Vanajal
119.	*Clerodendrum phlomidis*	Arni.	–	Arni	Agnimantha
120.	*Clerodendrum serratum*	Bharangi	–	Bharangi	Barangi
121.	*Clerodendrum viscosum*	Titabhamt	Hill Glory Bower	Titabha mt	Bhandirah
122.	*Clitoria ternatea*	Visnukanta	Butterfly pea	Aparajit a	Aparajita
123.	*Coccinia cord folia*	Kanduri	Little gourd	Kanduri	Bimbi
124.	*Cocculus hirsutus*	Jamtikibel	Broom creeper	Jaljamni	Patalgarudah
125.	*Cocus nucifera*	Nariyal	Coconut tree	Nariyal	Narikelah
126.	*Coffea arabica*	Kaphi	Coffee	Caphi	Rajapiluh
127.	*Coix lacryma – jobi*	Motida	Job's tears	Sankru	Gavedhukah
128.	*Coleus aromaticus*	Patthar chur	Indian borage	Patthar chur	Ashmabheda
129.	*Colocasia esculenta*	Arvi	Cocoyam	Arvi	Alupam
130.	*Commelina benghalensis*	–	Day flower	Kankau oa	Kanchata
131.	*Convolvulus arvensis*	–	Dear's foot	Harin padi	Rajbali

132.	*Convolvulus pluricauli*	Shankpushpi	–	Shankpu shpi	Shankpushpi
133.	*Corchorus olitorius*	–	Jew's mallow	Koshta	Pattshak
134.	*Cordia dichotoma*	Gunda	Sebesten plum	Lasura	Bahuvara
135.	*Coriandrum sativum*	Dhania	Coriander	Dhaniya h	Dhanaka
136.	*Costus speciosus*	Kevuk kand	–	Keu	Kebuka
137.	*Crinum defixum*	Vishaman Dala	–	–	Vishamandala
138.	*Crotalaria retusa*	Ghunghunia	Rattlewort	Ghunghu nia	Sanapuspi
139.	*Cryptolepis buchanani*	Kala bel	Wax Leaved Climber	Kala bel	Krishnasariva
140.	*Cucumis sativus*	Khera	Cucumber	Khira	Trapusah
141.	*Cucumis trigonus*	*Ban kachria*	Bitter gourd	Jangli Indrayan	Indravaruni
142.	*Cucurbita maxima*	Kaddu	Red gourd pumpkin	Kaddu	Pitakusmand ah
143.	*Cucurbita pepo*	Kumrha	Pumpkin	Kumrha	Kushmanda
144.	*Cuminum cyminum*	Jeera	Cumin	Jira	Jiraka
145.	*Curculigo orchioides*	Kali musli	Black musali	Musali	Talamuli
146.	*Curcuma amada*	Ambi haldi.	Mango ginger	Ama haldi	Amragandhi-haridra
147.	*Curcuma longa*	Haldi	Turmeric	Haldi	Haridra
148.	*Curcuma zedoaria*	Jangli Haldi	White turmeric	Jangli Haldi	Kachur
149.	*Cuscuta reflexa*	Amarbel	Dodder	Amarbel	Akasavalli
150.	*Cyamopsis tetragonoloba*	Guar	Cluster bean	Govar	Ksudrasimbi
151.	*Cymbopogon jwarancusa*	Bur	Rosha Grass	Rosha ghans	Jwarankush
152.	*Cymbopogon martini*	Rosha	Rosha grass	Gandha bel	Rohisah
153.	*Cynodon dactylon*	Dub	Doob grass	Doob	Niladurva
154.	*Cyperus esculentus*	–	Rush nut	Cicada	Bhandramusta
155.	*Cyperus rotundus*	Motha	Nut grass	Nagarmot ha	Musta
156.	*Dalbergia sissoo*	Shisham	Indian red wood	Shisham	Simsapa
157.	*Datura metal*	Kala dhatura	Thorn apple	Kala dhatura	Dhustura
158.	*Datura stramonium*	Safed dhatura.	Thorn apple	Safed dhatura	Dhattura
159.	*Delonix regia*	Gulmohar	Flame Tree,	Gulmohar	–
160.	*Dendrocalamus stirctus*	*Bans*	Solid bamboo	Bans	Vansha
161.	*Desmodium gangeticum*	Sarivan.	–	Salpan	Salaparni

162.	*Desmodium triflorum*	–	–	Kudaliya	Tripadi
163.	*Dichrostachys cinerea*	Bilatri, Sagun	–	Kheri	Vellantarah
164.	*Digera arvensis*	*Lat mahuria*	–	Latmahuria	Aranya vastuka
165.	*Dioscorea bulbifera*	Varahi kand	Patato yam	Ratalu	Varahi
166.	*Diospyros malabarica*	Kala tandu	Indian persimon	Tedu	Tinduki
167.	*Diospyros melanoxylon*	*Biripatta*	Coromandel ebony persimmon	Tendu	Dirghpatraka
168.	*Diplocyclos palmatus*	Ban-kakra	–	Shivlingi	Lingini
169.	*Dolichandrone falcata*	Medhshingi	-	Medhshingi	Mesasrnga
170.	*Dregea volubilis*	Akad bel	Cotton milk plant	Akad bel	Bahuparni
171.	*Drypetes roxburghii*	Putranjiva	Wid olive	Putranjiva	Putrajivah
172.	*Echinops echinatus*	Ontkata	Thistle	Ontkatara	Utkantaka
173.	*Eclipta prostrate*	Bhangra	Trailing eclipta	Bhangra	Bhringraj
174.	*Embelia tsjeriam – cottam*	Bayabirang	Basal	Bayabirang	Birangbhaida
175.	*Emilia sonchifolia*	–	–	Hirankhuri	Sasasrutih
176.	*Enicostemma littorale*	Nau	–	Chota chirayata	Nagajihva
177.	*Erythrina variegata*	Paribhadra	Indian coral tree	Dadap	Paribhadra
178.	*Eulophia nuda*	Amarkand	Spectacular Eulophia	Amarkand	Balakanda
179.	*Euphorbia antiquorum* L.	Tridhara	Triangular spruge	Tidhara	Vajrakantakah
180.	*Euphorbia hirta*	Dudhi	Australian asthma herb	Dudhi	Dugdhika
181.	*Euphorbia neriifolia*	Pattonkisend	Holy milk hedge	Shend	Sehund
182.	*Euphorbia nivulia*	Katathohar	Leafy Milk Hedge	Katathohar	Adhoguda
183.	*Euphorbia thymifolia*	–	Didiya	Dudhi	Dugdhika
184.	*Euphorbia tirucalli*	Thor	Milk bush	Sehund	Bahukshira
185.	*Evolvulus alsinoides*	Nil Shankhpuspi	–	Shyama kranta	Laghu-vishnuk-ranta
186.	*Exacum bicolor*	Avachiretta	Bicolor Persian Violet	Avachiretta	–
187.	*Ficus arnottiana*	*Paras-pipal*	–	Paras pipal	Plaksha
188.	*Ficus benghalensis*	Bargad	Banyan tree	Bargad	Vata

189.	*Ficus carica*	Anjeer	Common Fig	Anjeer	–
190.	*Ficus glomerata*	Gular	–	Gular	Udambara
191.	*Ficus microcarpa*	–	–	Kamarup	Plaksah
192.	*Ficus religiosa*	Pipal	Pipal tree	Pipal	Pippala
193.	*Flacourtia indica*	Bilanga	Madagasc ar plum	Kancu	Aghori
194.	*Flacourtia jangomas*	*Panialah*	Many spiked flacorita	Talispatri	Prachinamala ka
195.	*Foeniculum vulgare*	*Saunf*	Fennel	Saunph	Madhurika
196.	*Fumaria indica*	Pitpapra	–	Pitpapra	Parpat
197.	*Gardenia gummifera*	Chhote kurra			
198.	*Garuga pinnata*	*Kekad*	Garuga	Khogar	Krsnamlika
199.	*Glinus oppositifolius*	Jima	–	Jima	Lonika
200.	*Gloriosa superba*	Kalihari	Malabar glory lily	Karihari	Kalihari
201.	*Gmelina arborea*	Kamhar	White teak	Gamhar	Kasamari
202.	*Gossypium arboreum*	–	Tree cotton	Kapas	Raksatika
203.	*Grewia subinaequalis*	Phalsa.	–	Dhamin	Alpasthi
204.	*Grewia tenax*	Gondni	White Crossberry	Gondni	–
205.	*Grewia tilifolia*	*Dhaman*	Dhaman	Dhaman	Dhanva-ngah
206.	*Gymnema sylvestre*	Gurmar	Small Indian Ipecacuanha	Gurmar	Ajaballi, Mesasringi
207.	*Haldina cordifolia*	Haldu	Haldu	Haldu	Haridru
208.	*Hardwickia pinnata*	Anjan	Anjan	Anjan	–
209.	*Hedychium coronarium*	Dolan champa	Butterfly Ginger Lily	Dolan champa	–
210.	*Hedyotis herbacea*	–	–	Paper-bhed	Chayaparpatika
211.	*Helianthus annus*	–	Sunflower	Surajmu khi	Adityabhakta
212.	*Helicteres isora*	Marod phalli	East Indian screw tree	Bhendu	Avartani
213.	*Heliotropium indicum*	*Hathisunda*	Indian turnsole	Siriyari	Vrscikali
214.	*Hemidesmus indicus*	Anantmul	Indian sarsaparilla	Salsa	Sariva
215.	*Hibiscus cannabinus*	*Patsan*	Mesta	Pitwa	Anvastha
216.	*Hibiscus rosa-sinensis*	Gurhal	China rose	Gurhal	Japa
217.	*Hiptage benghalensis*	Madhavilata	Clustered hiptage	Madhu malati	Madhavi
218.	*Holarrhena pubescens*	Kurchi	Easter tree	Kurchi	Kutaja
219.	*Holoptelea integrifolia*	Chilbil	Indian elm	Papri	Chirabilva

220.	*Holostemma ada-kodien*	*Chhirvel*	–	Chirvel	Jivanti
221.	*Homonoia riparia*	*Surra*	–	Jalbent	Jalavetasah
222.	*Hordeum vulgare*	–	Barley	Jau	Yavah
223.	*Hybanthus enneaspermus*	Ratan purush	Spade Flower	Ratan purush	Rathnapurus ha
224.	*Hygrophila auriculata*	Talmakhana	–	Gokulk-anta	Vajrak-antaka
225.	*Hyptis suaveolens*	Vilaiti tulsi	American Mint	Vilaiti tulsi	Bhustrena
226.	*Impatiens balsamina*	Gul-mehndi	Garden Balsam	Gul-mehndi	Tairini
227.	*Indigofera tinctoria*	Neel	Indian indigo	Nil	Nilini
228.	*Ipomoea batatas*	Shakarkand	Sweet potato	Sakkark and	Pindaluh
229.	*Ipomoea nil*	Kaladana	Morning glory	Kaladan a	Krsnabija
230.	*Ipomoea sepiaria*	–	–	Bankal mi	Laksmana
231.	*Ixora coccinea*	–	Jungle-flame ixora	Rangan	Paranti
232.	*Jasminum arborescens*	*Chameli*	Tree jasmine	Bela	Vasanti
233.	*Jasminum auriculatum*	*Juhi*	Needle flower Jasmine	Juhi	Yuthika
234.	*Jasminum grandiflorum*	Chameli	Common jasmin	Cameli	Malati
235.	*Jasminum multiflorum*	Safed Chameli	Downy jasmine	Kundph ul	Kundah
236.	*Jasminm sambac*	Mogra	Arabian jasmin	Mogra	Mallika
237.	*Jatropha curcas*	Ratanjot	Purging nut	Ratanjot	Suklaeranda
238.	*Jatropha gossypifolia*	Lal – erenda	–	Lal – erenda	–
239.	*Justicia gendarussa*	Nili-Nargandi	–	Nilinirg undi	Vataghni
240.	*Justicia procumbens*	–	Water Willow	–	–
241.	*Kalanchoe pinnata*	Pattharchur	Sprout leaf	Zahhm-e-hayat	Asmabheda
242.	*Kyllinga nemoralis*	–	–	Nirbisi	Svetanirvisa
243.	*Lablab purpureus*	Sem	Lablab bean	Sem	Nispavah
244.	*Lagenaria siceraria*	Lauki	Bitter bottle gourd	Kadvi lauki	Katutumbi
245.	*Lagerstroemia speciosa*	–	Pride of India	Jarul	Tinisa
246.	*Lannea coromandelica*	Mohin	Indian Ash Tree	Mohin	Jinginee

247.	*Lantana camara*	–	Wild sage	Aripple	–
248.	*Lawsonia inermis*	Mehndi	Henna	Mehndi	Medika
249.	*Lens culinaris*	Masur	Lentil	Masur	Masurah
250.	*Lepidium sativum*	Halim	Water cress	Halim	Asalika
251.	*Leptadenia reticulata*	–	–	Dori	Jivanti
252.	*Leucas cephalotes*	Dhurpi sag	Tamba	Guma	Dronapuspi
253.	*Limnophilla aromatica*	*Kuttra*	–	Kutra	Amragandhah
254.	*Limonia acidissima*	Kabit	Elephant apple	Kait	Kapittha
255.	*Linum usitatissimum*	Alsi	Flax plant	Alsi	Atasi
256.	*Ludwigia octovalvis*	Ban long	Willow Primrose	Ban long	Bhu lavangah
257.	*Luffa acutangula*	Kadvi turoi	Bitter ribbed gourd	Kadvi turoi	Kosataki
258.	*Luffa cylindrica*	Ghia turoi	Sponge gourd	Khiya turoi	Mahakosataki
259.	*Macrotyloma uniflorum*	–	Horse gram	Kulthi	Kutatthah
260.	*Madhuca indica*	Mahua	Butter tree	Mahua	Madhuka
261.	*Malaxis acuminata*	–	Jeevak	Jivak	Jivakah
262.	*Mallotus philippensis*	–	Kamala tree	Kamala	Recanakah
263.	*Mangifera indica*	Aam	Mango	Aam	Amra
264.	*Manihot esculenta*	Cassava	Tapioca	Sakkark and	Kalapakandah
265.	*Martynia annua*	Bicchu	Devil's claw	Bichua	Kaknasa
266.	*Melia azedarach*	Bakayan	Persian lilac	Bakayan	Mahanimbah
267.	*Mentha spicata*	Podina	Garden mint	Podina	Putiha
268.	*Merremia emarginata*	–	–	Musakari	Akhukarni
269.	*Merremia tridentata*	–	–	–	Prasarani
270.	*Michelia champaca*	Champa	Champak	Campa	Campakah
271.	*Mimosa pudica*	Chuimui	Sensitive plant	Lajwanti	Lajjalu
272.	*Mimusops elengi*	Molsari	Bullet wood	Mulsari	Bakulah
273.	*Mimusops hexandra*	Khirni	Rayan	Khirni	Kshirini
274.	*Mirabilis jalapa*	Gulabbas	Four O'clock plant	Gulabasa	Trisandhi
275.	*Mitragyna parvifolia*	Karam	–	Karam	Kadambi
276.	*Momordica charantia*	Karela.	Bitter gourd	Karela	Karavelli
277.	*Momordica dioica*	Jangli Karela	Spiny gourd	Kakra	Karkotaki
278.	*Morinda pubescens*	–	Morinda tree	Acc	Paphanah

279.	*Moringa oleifera*	Sahjan	Horse-radish tree	Sahijan	Sobhan-janah
280.	*Morus alba*	Sehtut	Mulberry	Sehtut	Tutam
281.	*Mucuna pruriens*	Kemuch	Cowhage	Kaunc	Kapikacchuh
282.	*Mukia madraspatana*	Bilari	–	Bilari	Trikosaki
283.	*Murraya koenigii*	Kathrim	Curry leaf tree	Mitha nim	Kaidarya
284.	*Murraya paniculata*	Kamini	Chinese boxwood	Kamini	–
285.	*Naravelia zeylanica*	–	–	–	Dhanavalli
286.	*Nelumbo nucifera*	Kamal	Indian lotus	Kamal	Kamala
287.	*Nerium indicum*	Kaner	Indian oleander	Kaner	Karavira
288.	*Nervilia aragoana*	–	–	Sthalapad ma	Padmacarini
289.	*Nicotiana tabacum*	Tambaku	Tobacco	Tamaku	Tamakhuh
290.	*Nyctanthes arbor-tristis*	Harsinghar	Night jasmine	Harsing har	Sephali
291.	*Nymphaea nouchali*	Neelkamal	Indian red water lily	Koka	Kumuda
292.	*Ocimum basilicum*	Marwa	Sweet basil	Babui tulsi	Barbari
293.	*Ocimum canum*	Jungli tulsi	Sacred basil	Jungli tulsi	Arjaka
294.	*Ocimum gratissimum*	Ramtulasi	Shrubby basil	Bantulasi	Vanabarbarika
295.	*Ocimum kilimandscharicum*	*Kapur-tulsi*	Camphor basil	Kapurtu lsi	Karpuratulasi
296.	*Ocimum sanctum*	Tulsi	Holy basil	Tulsi	Tulasi
297.	*Oldenlandia corymbosa*	Dhaman-paper	–	Damanp apar	Parpatah
298.	*Operculina turpethum*	–	Indian jalap	Nisoth	Trivrt
299.	*Opuntia elatior*	*Naagphani*	Prickly pear	Naagaph anaa	Kantaseva
300.	*Oroxylum indicum*	Arlu	Indian trumpet tree	Arlu	Syonakah
301.	*Oryza sativa*	*Chawal*	Paddy	Chawal	Syali
302.	*Ougeinia oojeinensis*	*Tinsa*	Sandan	Sandan	Tinisah
303.	*Oxalis corniculata*	Amrulsak	Indian sorrel	Amrul	Cangeri
304.	*Oxystelma esculentum*	*Dudhiyalata*	–	Dudlata	Dughika
305.	*Pandanus fascicularis*	Kevda	Screw pine	Kevda	Ketaki
306.	*Panicum sumatrense*	–	Little millet	Savan	Syamakah
307.	*Paspalum scrobiculatum*	*Kodra*	Punctured paspalum	Kodo	Hudala
308.	*Pergularia daemia*	Sadowani	–	Utaran	Kurutakah
309.	*Phaseolus mungo*	*Urd*	Black gram	Urid	Masha

310.	*Phaseolus trilobatus*	Mugam	Green gram	Mooga	Mudgaparni
311.	*Phoenix dactylifera*	Khajur	Date palm	Khajur	Kharjurah
312.	*Phoenix sylvestris*	Khajuri	Date sugar palm	Khajuri	Kharjuri
313.	*Phyla nodiflora*	Jalpippali	Purple lippia	Jalpippali	Jalapippali
314.	*Phyllanthus emblica*	Amla	Emblic myrobalan	Amvala	Amalaki
315.	*Phyllanthus niruri*	Bhuiamla	Country gooseberry	Bhonya bali	Bhudhatri
316.	*Phyllanthus reticulatus*	–	–	Pancoli	Krsnakamboji
317.	*Physalis minima*	Ban phutka	Country gooseberry	Cirpoti	Cirapotha
318.	*Piper betle*	Paan	Betel	Pan	Tambulah
319.	*Piper longum*	Pipram	Indian long pepper	Pipli	Pippali
320.	*Pisum sativum*	Matar	Garden pea	Mathar	Kalayah
321.	*Plumbago zeylanica*	Chitrak	Ceylon leadwort	Chita	Chitrak
322.	*Plumeria rubra*	–	Pogodo tree	Cameli	Svetacampaka
323.	*Pongamia pinnata*	Karanj	Indian beech	Karanj	Karanjah
324.	*Populus alba*	–	–	–	–
325.	*Portulaca oleracea*	Kulfa	Common purslane	Badi loni	Brhat loni
326.	*Portulaca quadrifida*	Loniya	–	Chotalu niya	Laghughonika
327.	*Psidium guajava*	Amrud	Common guava	Amrud	Perukah
328.	*Psoralea corylifolia*	Babchi	Babchi seeds	Bakuchi	Bahuchi
329.	*Pterocarpus marsupium*	Bijasal	Indian kino tree	Bijasal	Bijaka
330.	*Pueraria tuberosa*	Patal kumhra	Indian kudzu	Vidarika nd	Vidari
331.	*Punica granatum*	Anar	Pomegranate	Anar	Dadimah
332.	*Quisqualis indica*	–	Rangoon creeper	Rangun achi veli	–
333.	*Randia dumetorum*	Mainphal	Emetic nut	Mainphal	Madana
334.	*Ranunculus sceleratus*	*Jaldhania*	Blister buttercup	Jal dhania	Kandira
335.	*Raphanus sativus*	Muli	Radish	Muli	Mulika
336.	*Rhus parviflora*	Khatua	–	Tung	Tintidika
337.	*Ricinus communis*	Arandi	Castor	Erandi	Svetaeranda
338.	*Rotula aquatica*	*Dantratchu*	–	–	Pasanabhedah
339.	*Rumex crispus*	Ambavati	Toothed Dock	Ambavati	Changeri

340.	*Rumex vesicarius*	Chooka	Bladder Dock	Chooka	Amlavetasa
341.	*Rungia repens*	Kharmor	–	Kharmor	Parpatha
342.	*Saccharum arundinaceum*	–	Devil sugarcane	Ramsar	Munjah
343.	*Saccharum munja*	–	–	Munj	Sara
344.	*Saccharum spontaneum*	–	Thatch grass	Kus	Kasah
345.	*Salvadora persica* L. Var. *Wightiana*	–	*Tooth brush tree*	Jhak	Angahava
346.	*Santalum album*	Chandan	Sandal wood	Safed chandan	Chandanam
347.	*Sapindus laurifolia*	Reetha.	Soapnut tree	Ritta	Phenilah
348.	*Saraca asoca*	Ashok.	Ashoka tree	Asok	Asoka
349.	*Sarcostemma acidum*	Somlata	Moon plant	Somlata	Somavalli
350.	*Schleichera oleosa*	Kusum.	Lac tree	Kusum	Mukulakah
351.	*Schrebera swietenioides*	Banpalash	Weaver's beam tree	Moka	Muskakah
352.	*Scirpus grossus*	–	–	–	–
353.	*Securinega leucopyrus*	Hartho	–	Aints	Bhuriphali
354.	*Semecarpus anacardium*	*Bhelwa*	Marking nut tree	Bhilawa	Bhallataka
355.	*Senna absus*	Chaksu	–	Chaksu	kulatthika
356.	*Senna auriculata*	Amoli	Tanner's cassia	Tarwar	Avartaki
357.	*Senna occidentalis*	Ban-chironta	Negro coffee	Kasmudi	Kasamardah
358.	*Senna tora*	Chakunda	Foetid cassia	Chakun da	Cakramardah
359.	*Sesbania grandiflora*	Agastya	Sesban	Agasti	Agasti
360.	*Sesamum indicum*	*Til*	Gingelly	Til	Tilah
361.	*Setaria italica*	–	Deccan grass	Bartiya	Kanghuni
362.	*Shorea robusta*	Sal	Sal tree	Sal	Agnivallabha
363.	*Sida acuta*	Baraira	Common Wireweed	Baraira	Bala
364.	*Sida cordata*	Bhuinii	Long-stalk Sida	Bhuinii	Bhumibala
365.	*Sida cordifolia*	Bala	Country mallow	Bariara	Bala
366.	*Sida rhombifolia*	Atibala	Common bale	Lal berela	Mahabala
367.	*Sida spinosa*	–	Prickly Sida	Gangern	Bala
368.	*Sisymbrium irio*	–	London-Rocke	Khubkalan	Khubakala
369.	*Solanum dulcamara*	–	Bitter sweet	–	Kakmachi
370.	*Solanum indicum*	Bruhati	Poison berry	Barhanta	Brhati

371.	*Solanum nigrum*	Makoi	Black night shade	Makol	Kakamaci
372.	*Solanum surattense*	Bhuri rindi	Yellow berries	Kateli	Kantakari
373.	*Soymida febrifuga*	Rohan	Indian redwood tree	Rohan	Manasrohini
374.	*Spermacoce hispida*	–	–	–	–
375.	*Sphaeranthus indicus*	Gorakhmundi	East Indian Globe Thistle	Gorakh mundi	Mahamundi
376.	*Sterospermum chelonoides*	*Padar*	Trumpet flower	Padhal	Alivallabha
377.	*Stereospermum colais*	*Chhota padar*	Trumpet flower tree	Padaria	Aliballabha
378.	*Strychnos potatorum*	*Kaya*	Celaring Nut	Nirmali	Katakah
379.	*Symplocos racemosa*	Lodhra	Lodh tree	Lodh	Lodhra
380.	*Syzygium cumini*	Jamun	Black plum	Jamun	Jambula
381.	*Tabernaemontana divaricata*	*Chandni*	East India Rosebay	Chandni	Tagar
382.	*Tabernaemontana heyneana*	–	East Indian rose bay	–	Sveta-kutaja
383.	*Tamarindus indica*	Imli	Tamarind	Imli	Amlika
384.	*Tectona grandis*	Sagwan	Teak	Sagun	Saka
385.	*Tephrosia purpurea*	Sarphonka	Wild indigo	Jungli nil	Sharapunkhaa
386.	*Teramnus labilis*	Mashoni	Wild udad	Masparni	Ardramasha
387.	*Terminalia arjuna*	Arjun	Arjuna	Arjun	Arjuna
388.	*Terminalia bellerica*	Bahera	Beleric myrobalan	Bahera	Bibhitaka
389.	*Terminalia chebula*	Harra	Black myrobalan	Harra	Haritaki
390.	*Terminalia crenulata*	Saj	Black murdah	Asaina	Chagakarnah
391.	*Thalictrum foliolosum*	Pilzari	–	Pilazari	Mamira
392.	*Thespesia populnea*	–	Portia tree	Paraspip al	Haripucchah
393.	*Thevetia peruviana*	Pili kaner	Yellow oleander	Pili kaner	Pitapuspa
394.	*Tinospora cordifolia*	Gurcha	Heart leaved moonseed	Amrita	Guduchi
395.	*Toona ciliata*	Tuna	Red cedar	Tuna	Apina
396.	*Trachyspermum ammi*	Ajwain	Carum	Ajwain	Ajawodika
397.	*Trapa bispinosa*	Singhara	Water chestnut	Singaad ha	Srngataka
398.	*Trianthema portulacastrum*	–	Horse purslane	Sabuni	Chiratika
399.	*Tribulus terrestris*	Gokhru	Land caltrops	Gokhru	Goksura
400.	*Trichodesma indicum*	Chhota-Kulpha	–	Chhota kulpa	Surasa

401.	*Trichosanthes cucumerina*	Jangli-chichinda	–	Jungli chachin da	Amritahala
402.	*Tridax procumbens*	–	Coat buttons	–	–
403.	*Trigonella foenum-graceum*	Methi	Fenugreek	Methee	Methika
404.	*Triticum aestivum*	Genhu	Bread wheat	Gehun	Arupa
405.	*Triumfetta rhomboidea*	–	Spiny cockle burr	Chikti	Jhinjharita
406.	*Typha elephantina*	Aer	Elephant grass	Mothitri na	Gundra
407.	*Uraria picta*	Dabra	–	Pithwan	Prishth-iparni
408.	*Urginea indica*	Jangli kanda	Indian squill	Jangli piyaz	Kolakanda
409.	*Urtica diocia*	Bichchhu	Stinging Nettle	Bichchhu	Vrscikali
410.	*Ventilago maderaspatana*	Keonti	Red creeper	Pitti	Raktavalli
411.	*Vernonia cinerea*	Sahadevi	Purple fleabane	Sahadevi	Sahadevi
412.	*Vetiveria zizanioides*	–	Vetiver	Khas-khas	Usira
413.	*Vigna radiata*	Moong	Green gram	Mung	Mudgah
414.	*Vigna radiata* (L.) Wilczek var. *Sublobata*	Jungli Moong	Wild black gram	Banurd	Masaparni
415.	*Vigna unguiculata*	Lobia	Cow pea	Sonta	Chavala
416.	*Vitex negundo*	Negur	Chaste tree	Shamalu	Nirgudi
417.	*Vitis vinifera*	*Angoor*	Grape vine	Angura	Draksa
418.	*Withania somnifera*	Ashwagandha	–	Ashwag andha	Ashwagandha
419.	*Woodfordia fruticosa*	Dhao	Fire flame	Dhai	Dhataki
420.	*Wrightia tinctoria*	Indrajau	Sweet indrajao	Indrajau	Indrayava
421.	*Xanthium strumarium*	Adhasisi	Burweed	Banokra	Arishta
422.	*Zingiber officinale*	Adrak	Ginger	Ada	Ardraka
423.	*Ziziphus mauritiana*	Jharberi	Indian plum	Ber	Badari
424.	*Ziziphus nummularia*		Wild jujube	Jharberi	Kshudrabadar
425.	*Ziziphus oenoplia*	Makoya	Jackal Jujube	Makai	Srigalakoli